Man and His Thermal Environment

Man and His Thermal Environment

R. P. Clark
Head of the Laboratory for Aerobiology,
Clinical Research Centre, Harrow, Middlesex

O. G. Edholm
Visiting Professor, Bartlett School of Architecture,
University College, London

Edward Arnold

© R.P. Clark and O.G. Edholm 1985

First published 1985 by
Edward Arnold (Publishers) Ltd,
41 Bedford Square, London WC1B 3DQ

British Library Cataloguing in Publication Data

Clark, R.P.
  Man and his thermal environment.
  1. Body temperature—Regulation
  I. Title   II. Edholm, O.G.
  612′.01426   QP135
  ISBN 0-7131-4445-9

Library of Congress Catalog Card Number 84-71792

Text set in 10/11pt Monophoto Times
and printed and bound in Great Britain by
Butler & Tanner Ltd, Frome and London

Preface

Man in a Cold Environment by A. Burton and O.G. Edholm was published in 1955. The present work may be regarded as a descendant; it is not a new edition but a new book, although it owes much to its predecessor. In the period of 30 years since the writing was completed there have been many developments and a greater amount of interest in the topic and this has resulted in many publications dealing with temperature regulation in animals and man. Yet another may seem an unnecessary venture, but this book, to some extent, represents the work of the Division of Human Physiology at the National Institute for Medical Research and emphasizes the developing knowledge of heat exchange between man and his environment. One of us (OGE) has been associated with studies of man in different thermal conditions, specifically in the climatic chambers at the National Institute, and RPC began his studies on heat exchange with the late Harold Lewis, who was a senior member of the Division of Human Physiology.

There has been no attempt to review the extensive publications of recent years, as these are so well covered in a number of books. Outstanding is *Thermoreception and Temperature Regulation* by Herbert Hensel published in 1981 as a monograph for the Physiological Society. The death of Herbert Hensel in 1983 at a tragically early age has deprived the subject of a physiologist with a remarkable research output.

The wealth of work on animals is only briefly mentioned; as the title of this book emphasizes, human physiology is the central interest.

A feature of this book is the colour illustrations derived from infra-red thermographic studies, which show the way in which the skin temperature varies, especially in conditions such as exercise. The use of thermography has made it necessary for physiologists to revise their concepts of skin temperature and this topic is dealt with in some detail. There are substantial sections dealing with descriptions of heat loss by convection which are markedly different from those described in *Man in a Cold Environment*. The changes in our concepts have been largely due to the use of the Schlieren optical technique, first used in this context by the late Harold Lewis.

This work is aimed, first of all at physiologists, especially those involved in climatic, environmental and thermal physiology. Additionally, it should be of use to ecologists and environmentalists in general and possibly to agriculturalists. We hope that there will be some clinical interest, especially in relation to the description of the infant microclimate and the diagnostic uses of thermography which are new topics. Another group who might find the material of use include architects and air-conditioning engineers who would be able to follow, not only the recent physiological work on thermal comfort, but also apply the current knowledge to a solution of their own problems.

London 1984   RPC
                                                                    OGE

Acknowledgements

We are grateful to M.R. Goff, A.B.I.P.P., A.M.P.A., A.R.P.S. and B.J. Mullan, A.B.I.P.P. for the preparation of the colour thermographic and Schlieren photographs and to P. Smith of AGA Infra-red Systems Ltd, for funds enabling us to include the colour plates in this book. We also thank Miss Winifred Bailey for checking the text and Mrs Cynthia Richards for typing the manuscript.

Contents

1

The Thermal Environment

Temperature influences all forms of life and all substances. Although this had been known for a long time, at first rather imprecisely, as methods and techniques of measurement improved more exact relationships could be demonstrated, which led to statements or 'laws' applicable to gases, fluids and solids. Plants and animals are also subject to these physical laws but it has nevertheless taken some time for the ubiquity of temperature to be appreciated. Man is no exception, and it is the way that temperature relates to human life which is the subject of this book.

Definitions of temperature and heat transfer

One of the objects of this work is to describe the temperature distribution over the body surface, and how this in turn leads to an exchange of heat with the surroundings in order to maintain a thermal equilibrium between the body and the environment. An important preliminary to this is to define the terms 'temperature' and 'heat loss'. Temperature is a property of substances with which we are all familiar and we use the term freely about our surroundings, although an exact definition of the term is difficult. The most obvious way in which we become aware of temperature is as a sense of hotness or coldness when we touch an object. We also quickly learn what happens when a hot and cold body are brought into contact; the former becomes cooler as the latter become warmer. If these two bodies remain in contact for some time they generally appear to acquire the same temperature. Sometimes our sense of temperature can be misleading. This happens when very cold bodies may seem hot, or bodies of different materials which are at the same temperature appear to be dissimilar.

These are some of the difficulties in appreciating temperature and because of this we must define equality of temperature. Consider two blocks of copper, one hot and the other cold, with their temperatures registered by mercury thermometers; if these two blocks are brought into thermal contact it will be observed that the electrical resistance of the hot block decreases with time and that of the cold block increases. After a period, no further change in resistance is observed. At the instant the blocks are brought together the length of the side of the hot block decreases with time, whereas the cold block dimension increases. After some time no further change in the length of either block is observed. Similarly, the mercury in the thermometer of the hot block at first begins to drop and in the cold block it rises until eventually no further change is noted in the heights of the mercury column. We can therefore say that two bodies have *equality of temperature* when no change in an observable property occurs if they are in thermal communication.

This concept can now be used to investigate a law of thermodynamics, less well known than the first and second laws, called the Zeroth law. It states that when any two bodies are equal in temperature with a third body, they in turn have equality of temperature with each other. This may seem an obvious statement based upon our everyday experience; however, the fact cannot be derived from the first or second laws of thermodynamics and because it can logically be said to proceed from them it is called the Zeroth law. This may be illustrated by again considering two blocks of copper together with a mercury thermometer. Let one block be brought into contact with the thermometer until they have equal temperatures; if the block is then removed and the second block of copper is brought into contact with the thermometer with no resultant change in the mercury level of the thermometer during the procedure, then we may say that both blocks of copper are in thermal equilibrium with the given thermometer and that the two blocks themselves have the same temperature.

This law is the foundation for temperature measurement, as it enables numbers to be assigned to the mercury thermometer so that each time a body has equality of temperature with the thermometer it may be said that the body has a temperature indicated numerically on the thermometer.

There is often an interchange and misuse of the terms 'heat' and 'temperature' in the everyday sense of these words. It is therefore important to clearly define heat, since it is basic to so many thermodynamic problems. Heat-flow is an exchange of energy between bodies which themselves have different energy levels characterized by their temperatures. The direction of heat flow is always from the higher to the lower temperature. If a hot body (again for instance a block of hot copper) is placed in a vessel containing cold water, it is common experience that the copper will cool down and that the water will warm until the temperatures are equal. These changes are caused by the fact that energy is transferred from the copper to the water; this enables us to define heat as that form of energy transferred across a boundary by virtue of a temperature difference or gradient. Heat is only transferred when the copper is placed in the water and the copper and the water are in thermal contact. This transfer of heat lasts only until the temperatures of the copper and the water are equal. When this occurs, there is no longer a temperature difference between the two bodies and consequently there is no longer any heat transfer.

The degree of molecular activity of a substance confers on it an energy level that may be termed 'heat' although the expression 'internal energy' is often used for this and the word 'heat' reserved for the transfer of energy to or from the body. In this regard heat may be considered as a transient phenomenon which is only measured indirectly when a body gains or loses heat across a boundary.

Heat may be calculated in terms of how much is required to produce the effect of a specified temperature change in a given quantity of water. For instance the calorie is the amount of heat needed to increase the temperature of 1 g of water by 1°C; the Calorie or kilogram calorie (kcal) being 1000 times larger is the heat required to raise the temperature of 1 kg of water through 1°C. The British Thermal Unit is the heat necessary to raise the temperature of 1 lb of water by 1°F.

Table 1.1 gives a comparison between different units used in heat transfer calculations.

Table 1.1 This shows a comparison between different units used in heat transfer calculations.

1 calorie	= 4.184 J (approx. 4.2 J)
1 kilocalorie or Calorie	= 4184 J (approx. 4200 J)
1 kilowatt hour	= 3 600 000 J
1 British thermal unit	= 1055 J (approx. 10^3 J)
1 therm	= 10^5 BTU (approx. 10^8 J)

Because heat is considered as a form of energy it must be subject to the principle of the conservation of energy which states that 'energy cannot be created or destroyed but only converted from one form to another'. Heat may therefore be changed into, or derived from, other forms of energy, for instance kinetic, electrical, mechanical, chemical, etc.

Measurements of heat flow may be made from the skin surface (see Chapter 2) but require quite different techniques from those for measuring temperature. However heat flow can often be deduced from measurements of body and environmental temperatures.

Thermal capacity and specific heat

When equal masses of different substances are heated by the same source, the temperature rise of each may be quite different. For example, equal masses of aluminium and copper may accept the same amount of heat but the aluminium will show a smaller increase in temperature and will be described as having a greater thermal capacity than the copper. When thermal capacity is expressed in terms of unit mass and temperature change it is known as specific heat. This is defined as the heat required to raise 1 kg of a substance by 1°C.

Examples of the two specific heats of various substances are given in Table 1.2.

Table 1.2 Examples of the specific heats of various substances.

Substance	$J.kg^{-1}°C^{-1}$
Air	7.1×10^3
Aluminium	0.92×10^3
Copper	0.39×10^3
Glass	0.13×10^3
Ice	2.1×10^3
Water	4.2×10^3
Wood	17.0×10^3
Body tissue	3.5×10^3

Newton's law of cooling

This law, which is an approximation for warm substances that are cooling in well-ventilated areas, states that the rate at which a body loses heat is proportional to the excess temperature of the body over its surroundings. Although useful for calculations involving the cooling of 'passive' bodies it is not of great value in studies of human heat loss because the body is constantly producing metabolic heat energy to combat heat loss and maintain a constant excess temperature over the environment.

Hot and cold

What do we mean when we say that an object is hot; do we mean that it has a high surface temperature or that it has a large heat production and energy exchange with the environment? In Plate 15 the thermograms of an infant nursed in an incubator show skin temperatures over the head to be no higher than over the trunk and sometimes slightly lower and yet we know that the metabolism in the infant's brain is some 20 per cent of the total. There is clearly a very efficient energy exchange between the brain and the rest of the body with little direct conduction of heat energy through the skull. In this sitiation is the head hot or cold? In terms of energy production it could be said to be hot but as far as surface temperature is concerned it could be termed cold.

In conditions of high air movement in cool ambient environments those areas of the skin which have the lowest temperature may also have the highest heat exchange with the surroundings. A most important principle is that the surface temperature may be a consequence of the heat exchange process rather than the determining factor for the energy transfer. This principle is discussed further in relation to convective heat transfer and skin temperature distribution patterns in Chapter 4 and the insulation requirements of clothing for harsh environments in Chapter 10.

Laws of thermodynamics

When the human body exchanges heat with the environment it must do so in strict accordance with the laws of thermodynamics and in this respect it is no different from any isolated physical, chemical or biological process where heat exchange with the surroundings is concerned. The Zeroth law has already been discussed and the laws which follow from it may be simply stated:

1. Heat and mechanical work are mutually convertible and in any process involving such conversion, one calorie of heat is equivalent to 4.18×10^7 ergs of mechanical work.
2. Heat cannot be transferred by any continuous, self-sustaining process from a cold to a hotter body.

A balance between the work done by the body, its metabolism and exchange of heat with the environment is required by the first law of thermodynamics and may be expressed by the following basic equation:

$$M - W = E + C + R + K + S$$

where M is the metabolism or rate of heat production, W is the useful mechanical work, E the heat lost by evaporation, C the heat exchanged by convection and R that exchanged by radiation. The heat loss by conduction is K and S the rate of body heat storage or change of body heat capacity which may either be increased or depleted to result in higher or lower body temperatures respectively.

Heat storage

In the above equation for human thermal balance, the heat storage term S can, in certain circumstances, be an important factor, particularly in the short term where in exercise, for instance, increased energy production may not be entirely matched by environmental cooling. In this case increased heat storage will result in a raised body temperature. If body temperature changes by 1°C/hour the rate of energy storage or depletion is some 68 watts or about 38 watts/m² of body area. The formula for evaluating heat storage is:

$$S = mC_p \frac{\dot{T}_d}{A_d}$$

where m is the body mass, C_p the specific heat of the tissues, $\dot{T}_d$ the rate of change of mean body temperature and A_d the skin surface area.

Children and infants are less able to store heat energy than adults because of their small body mass and low mass to surface area ratio. They are therefore particularly vulnerable in cold conditions to a lowering of body temperature. In the case of premature babies nursed naked in incubators precise environmental control is critical and some of the problems associated with this are discussed in Chapter 11.

Human metabolism

The various components of the energy balance equation can have a wide range of values depending on individual energy expenditure and environmental conditions. It is however possible to compare metabolic rates between individuals by an index known as the basal metabolic rate (BMR). This is the rate of energy production by the body during absolute rest determined under specific conditions by the rate of oxygen usage. The BMR is then expressed as the energy expenditure per unit surface area per hour. Figure 1.1 shows the variation of metabolic rate with age for males and females; BMR in the newborn infant in relation to body surface area is approximately twice that of a very old person. Children also have high cellular reaction rates as well as rapid synthesis of cells in the growing body.

Different activities markedly affect energy expenditure and Table 1.3 illustrates some relationships between work rate and energy production and shows a 20-fold difference between sleeping and walking upstairs.

Sex and growth hormones together with climate and sleep can all produce

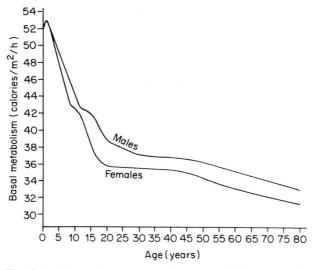

Fig. 1.1   Normal basal metabolic rates at different ages for each sex (from Guyton (1977) *Text book of Medical Physiology.* W. B. Saunders).

Table 1.3   Energy expenditure during different activities for a 70 kg man.

Activity	Energy expenditure watts (J/s)
Sleeping	80
Awake—lying still	90
Sitting at rest	115
Standing relaxed	120
Typing rapidly	160
'Light' exercise	200
Walking slowly (2.6 m.p.h.)	230
'Active' exercise	340
'Severe' exercise	520
Sawing wood	560
Jogging (5.3 m.p.h.)	660
Athlete—marathon running	1400
Athlete—sprint running	1750

changes in BMR of between 10 and 20 per cent but in fever the metabolic rate may be increased by as much as 100 per cent.

The secretion of thyroxin by the thyroid gland may increase metabolism by up to 100 per cent by direct action on the chemical reaction rates within the cells. A hyperthyroid person may typically have a BMR 40–80 per cent higher than normal and in hypothyroidism it may be reduced by 40–50 per cent.

The constancy and relatively small range of body temperature are discussed elsewhere. If body heat storage, manifested by increased temperature, is not to endanger the body in cases of increased metabolism, it is necessary for environmental heat exchange to be able to adjust to these conditions. In this respect the skin is a remarkable organ; it is the interface with the environment and is where tissue conductance, sweating and skin blood flow interact with the temperature, humidity and air movement of the surroundings to produce a thermal balance for comfort and ultimately for survival.

Body surface area

The various environmental and physiological processes involved in each term of the body heat balance equation are complex as are many of the experimental methods for their measurement which are outlined throughout this book. Heat transfer is usually evaluated in terms of coefficients where it is necessary to know the surface area of the body in order to calculate total heat loss.

It is difficult to determine precisely the surface area and the following formula was proposed by DuBois and DuBois[1] relating the area in m^2 (A) to the height (H) cm and weight (W) in kg.

$$A = 0.00718 \, W^{0.425} \times H^{0.725}$$

Although this formula has been found to underestimate the area by up to 7 per cent (Van Graan[2]) and various improvements have been proposed, it is adequate for most purposes and is presented in a convenient form in Table 1.4.

Temperature scales

In 1595, Galileo invented a simple gas thermometer known as a 'thermoscope' which, although unreliable, was used by Sanctorius of Padua to collect information about temperatures in hospital patients. Glass bulb thermometers containing alcohol appeared some 50 years later but did not achieve importance in medicine or science during that period. In 1714 Fahrenheit in Danzig developed the mercury in glass thermometer and defined the scale that now bears his name. Sometime later the scale was slightly revised to be fixed in terms of the ice and steam points and the 'normal' for the human body was then found to be 98.4°F (oral temperature).

Ice point is defined as the temperature of a mixture of water and ice which is in equilibrium with saturated air at a pressure of one atmosphere. The steam point is the temperature at which steam and water are in equilibrium at a pressure of one atmosphere. In Fahrenheit notation they are assigned numbers 32 and 212.

Although useful in medical work the Fahrenheit scale was not ideally suited to scientific use and in 1742 Celsius, the Swedish astronomer, proposed a scale with 100 degrees between the freezing and boiling points of water which became known as the Centigrade scale. In the early part of the nineteenth century greater precision in temperature measurement was again sought which led Henry Regnault in France to develop gas thermometers accurate to 0.1°C. Regnault also determined the expansion rate of hydrogen at

Table 1.4 Surface areas at various heights and weights calculated according to the DuBois Formula.

Height (cm) → Weight (kg) ↓	90	95	100	105	110	115	120	125	130	135	140	145	150	155	160	165	170	175	180	185	190	195
10	0.50	0.52	0.54	0.56																		
12.5	0.55	0.57	0.59	0.61	0.64																	
15	0.59	0.62	0.64	0.66	0.69	0.71	0.73															
17.5	0.63	0.66	0.68	0.71	0.73	0.76	0.78	0.80														
20	0.67	0.70	0.72	0.75	0.78	0.80	0.83	0.85	0.88	0.90												
22.5			0.76	0.79	0.82	0.84	0.87	0.89	0.92	0.95	0.97	1.00										
25				0.82	0.85	0.88	0.91	0.94	0.96	0.99	1.02	1.04	1.07									
27.5				0.86	0.89	0.92	0.95	0.97	1.00	1.03	1.06	1.08	1.11	1.14	1.16							
30					0.92	0.95	0.98	1.01	1.04	1.07	1.10	1.13	1.15	1.18	1.21	1.24						
32.5					0.95	0.98	1.02	1.05	1.08	1.11	1.14	1.16	1.19	1.22	1.25	1.28	1.31					
35						1.02	1.05	1.08	1.11	1.14	1.17	1.20	1.23	1.26	1.29	1.32	1.35					
37.5							1.08	1.11	1.14	1.17	1.21	1.24	1.27	1.30	1.33	1.36	1.39	1.42				
40								1.14	1.17	1.21	1.24	1.27	1.30	1.33	1.37	1.40	1.43	1.46				
42.5								1.17	1.21	1.24	1.27	1.30	1.34	1.37	1.40	1.43	1.46	1.50	1.53			
45									1.24	1.27	1.30	1.34	1.37	1.40	1.44	1.47	1.50	1.53	1.56			
47.5									1.26	1.30	1.33	1.37	1.40	1.44	1.47	1.50	1.53	1.57	1.60	1.63		

50	1.29	1.33	1.36	1.40	1.43	1.47	1.50	1.54	1.57	1.60	1.64	1.67	1.70
52.5		1.36	1.39	1.43	1.46	1.50	1.53	1.57	1.60	1.64	1.67	1.70	1.74 1.77
55		1.38	1.42	1.46	1.49	1.53	1.56	1.60	1.63	1.67	1.70	1.74	1.77 1.80
57.5			1.45	1.48	1.52	1.56	1.59	1.63	1.66	1.70	1.74	1.77	1.80 1.84
60			1.47	1.51	1.55	1.59	1.62	1.66	1.70	1.73	1.77	1.80	1.84 1.87
62.5				1.54	1.58	1.61	1.65	1.69	1.72	1.76	1.80	1.83	1.87 1.91
65				1.56	1.60	1.64	1.68	1.72	1.75	1.79	1.83	1.86	1.90 1.94
67.5					1.63	1.67	1.71	1.74	1.78	1.82	1.86	1.90	1.93 1.97
70					1.65	1.69	1.73	1.77	1.81	1.85	1.89	1.92	1.96 2.00
72.5						1.72	1.76	1.80	1.84	1.88	1.91	1.95	1.99 2.03
75						1.74	1.78	1.82	1.86	1.90	1.94	1.98	2.02 2.06
77.5							1.81	1.85	1.89	1.93	1.97	2.01	2.05 2.09
80							1.83	1.87	1.92	1.96	2.00	2.04	2.08 2.12
82.5								1.90	1.94	1.98	2.02	2.06	2.10 2.14
85								1.96	2.01	2.05		2.09	2.13 2.17
87.5								1.99	2.03	2.07		2.12	2.16 2.20
90									2.06	2.10		2.14	2.18 2.22
92.5									2.08	2.12		2.17	2.21 2.25
95										2.15		2.19	2.23 2.28
97.5										2.17		2.22	2.26 2.30
100												2.24	2.28 2.33
102.5												2.26	2.31 2.35
105													2.33 2.38
107.5													2.40

approximately 0°C to be 1 part in 273. This work complemented that of Lord Kelvin in England and others who were able to define 'absolute zero' on the gas thermometer as −273° on the Celsius scale by considering the laws of thermodynamics applied to the heat exchanges in an ideal heat engine cycle. Kelvin was then able to devise an absolute temperature scale in which the 'triplepoint' of water (the temperature at which ice, water and water vapour can coexist) was equivalent to 0°C. This is now known as the Kelvin scale (K) where $273.15° K = 0°C$ and Figure 1.2 shows a comparison of the Kelvin, Celsius and Fahrenheit scales.

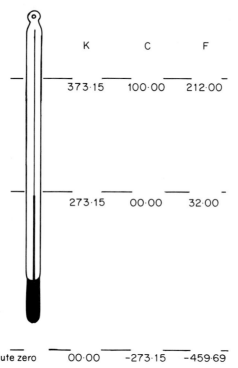

	K	C	F
	373·15	100·00	212·00
	273·15	00·00	32·00
Absolute zero	00·00	−273·15	−459·69

Fig. 1.2   Comparison of the Kelvin, Celsius and Fahrenheit temperature scales.

Temperature and metabolic rate

Mathematical expressions have long been sought to describe the effects of temperature on the metabolism of organisms and on the rate of chemical reactions. One of the first attempts was to define a temperature coefficient or Q_{10} value for these reactions. This is the ratio of the rate of a metabolic process (growth or microbial activity) at one temperature to the rate at a temperature 10°C lower. This can be expressed by:

$$Q_{10} = \frac{k_{(t+10)}}{k_t}$$

where k is the velocity constant of the reaction and t the temperature. The Q_{10} values were originally computed for chemical reactions but later they were applied to biological processes. For most metabolic processes the value of Q_{10} is between 2.0 and 3.0, i.e. heat production of living cells will increase 2-3 times if the temperature is raised by 10°C.

A disadvantage of a temperature coefficient for detailed studies of the effects of temperature of biological processes is that the values for any reaction vary for the span over which the rates are measured. A more general description of the effect of temperature on biological systems was given in 1899 by Arrhenius, who derived the following equation to describe the effect of temperature on rates of chemical reactions:

$$k = Ae^{-E/RT}$$

In this expression k is the velocity (heat production rate for instance) for the particular reaction and T is the temperature in degrees absolute. Since the terms A and E in this expression are usually assumed to be constant for a particular reaction, the plot of $\log_{10}k$ against $1/T$ should give a straight line with a slope of $-E/2.303R$. This relationship has been shown to hold for a wide range of chemical reactions although a number of biological processes, in particular microbial growth and enzymic reactions studied in vivo and in vitro, characteristically produce non-linear plots. Nevertheless, the Arrhenius relationship is useful for comparing the effects of temperature on different biological processes.

The Q_{10} relationship can be combined with the Arrhenius equation to give the following expression relating Q_{10} to the rate of the reaction at any two temperatures t_1 and t_2.

$$Q_{10} = \left(\frac{k_1}{k_2}\right)^{10/(t_1 - t_2)}$$

As an example, if the Q_{10} is 2.5 then a rise of 1°C increases heat production (k_1/k_2) by a factor of $2.5^{0.1}$ or about 9.6 per cent (the Q_{10} of 2.3 would give an increase of about 8.7 per cent for a 1°C rise).

Although the relationships described above have been formulated with regard to chemical or isolated biological processes, there have been a number of studies to determine whether the relationship holds good for the whole body in man or other homeotherms. In this case, the rate of reaction can be characterized by the rate of oxygen consumption (VO_2), and the reaction temperature is the deep body temperature. Measurements of VO_2 when body temperature is lowered in man and other mammals have produced extremely variable results. This is because mammals actively increase their heat production in an attempt to prevent the body temperature from falling, and this effect is not abolished until the body temperature has fallen well below 32°C. It is only when all of the thermoregulatory responses (shivering, changes in skin blood flow, etc.) to cooling have been eliminated that the metabolic response to body temperature alone can be isolated from the complete thermoregulatory 'control system'. This can be achieved by deep general anaesthesia or surgical procedures which damage the brain and this has been studied in a number of cases, again with variable results. Cross *et al.*[3] have summarized much of this work whilst at the same time describing a study in which they were able to investigate a newborn anencephalic baby which was hypothermic on delivery. This infant showed no increase in metabolic rate in response to environmental cooling as with all normal infants. During rewarming from a rectal temperature of 29°C to 37.5°C its oxygen consumption increased at a rate that produced a Q_{10} value of 2.1. It is perhaps remarkable that such a close correlation exists between temperature and metabolic rate for isolated chemical and biological reactions and for the whole body of homeotherms. More evidence for this is seen in Table 1.5 (from Cross *et al.*[3]) which gives values of Q_{10} from the results of a number of authors who reported studies on various mammals.

Table 1.5   Distribution of Q_{10} values—mammalian material.

Reference number	Source	Q_{10}	Range of body temperature studied (°C)	Animal
3	Cross *et. al.* (1966)†	2.1	37–29	Anencephalic infant in vivo
4	Himwich *et al.* (1940)	2.2	44–25	Adult rat-minced or sliced brain cortex in vitro
5	Field *et. al.* (1944)†	2.1	37–10	
6	Velten (1880)†	2.2	38–23	Adult rabbit in vivo
7	Krogh (1914)†	2.1	37–14	Young dog in vivo
8	Bigelow *et al.* (1950)	2.6	38–18	Adult dogs in vivo
9	Lynn *et al.* (1954)	2.5	38–18	Adult dogs in vivo
10	Spurr *et al.* (1954)†	2.35	37–23	Adult dogs in vivo
11	Deterling *et al.* (1955)	2.4	37–20	Adult dogs in vivo
12	Bickford & Mottram (1960)†	2.3	40–25	Adult rabbits in vivo
13	Civalero *et al.* (1962)	2.3	37–17	Adult dogs in vivo
14	Hockaday *et al.* (1962)†	2.3	37–34	Adult female patient

Mean Q_{10} value = 2.29 ± 0.16 (SD)
†   In papers marked thus the authors were of the opinion that their results accorded with a Q_{10} effect.

Figure 1.3 shows the relationship between the percentage increase in heat production (or metabolic rate) for a 1 per cent temperature rise at values of Q_{10} between 2 and 3. From this it may be seen that the precision of measurement of Q_{10} on the whole bodies of 'normal' adults relies heavily on accurately estimating the changes in metabolic rate resulting from changes in temperature.

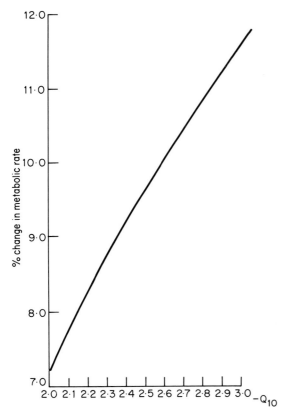

Fig. 1.3 Percentage increase in heat production for each 1 per cent temperature rise at values of Q_{10} between 2 and 3.

Temperature regulation

There will be many references to temperature regulation throughout this book, but no detailed description, as some excellent accounts have been published[15, 16, 17, 18] which it seems unnecessary to duplicate. Instead, a brief introduction follows and temperature regulation during exercise will be described.

Body temperature in man and other mammals remains relatively constant in spite of large changes in heat production and in widely differing environmental conditions. Such a constancy is achieved by an active regulation controlling the rate of heat loss, and to a lesser extent, the rate of heat production. The centre for this regulation is the hypothalamus, where blood temperature is sensed and where information is received from the temperature sensors in the skin. It is still uncertain if there are significant temperature sensors in any other regions of the body, although there is substantial evidence that the spinal cord is temperature sensitive and can act as a regulator if the hypothalamus is damaged. Within the hypothalamus, the input from the

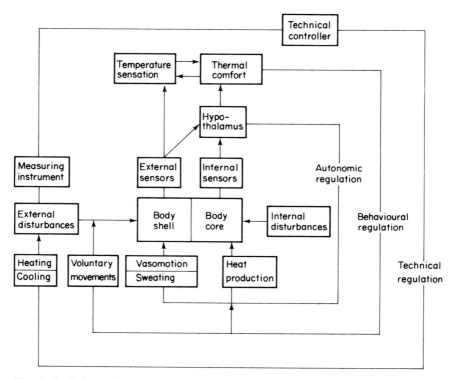

Fig. 1.4   Schematic diagram of autonomic, behavioural and technical temperature regulation in man (reproduced with permission from Hensel (1981)[17]).

temperature sensors is relayed to various cells, from where the output passes to centres controlling blood vessels, sweat glands and the muscular activity of shivering. As a result, heat loss from the skin is modified and heat production may also be changed. Skin temperature can be modified as the input to the hypothalamus is altered. There is a closed loop with a feed-back system, enabling control to be maintained automatically. Many diagrams have been prepared to illustrate this arrangement; an example is shown in Figure 1.4.

This simple description of the basic elements of temperature regulation is based on extensive experimental work by many investigators. There are a number of unanswered questions; the most obvious is the level of the control temperature and the mechanism needed for an appropriate reference temperature. The problem is illustrated by the fact that body temperature is not controlled at a contant level; there is a circadian rhythm with an amplitude of approximately 1°C. Even more remarkable is the regulation during exercise, when the increased heat production raises body temperature[19]. This rise is controlled in such a way that the body temperature increases proportionally to the level of exercise, and is unaffected between wide limits by the ambient temperature. This observation by Nielsen[20] has been confirmed many times, notably by Lind[21], but there is still lack of agreement about the physiological mechanisms involved.

Changes in skin temperature during exercise have been recorded using infra-red thermography and these are described in Chapter 3. They may be summarized by saying that there is vaso-constriction in the extremities and in regions not involved in the exercise. In contrast, the skin overlying muscle shows a rise in temperature but over the abdomen, for example, there is a fall. During running the skin of the thighs and legs becomes warm due to the heat production of the muscles beneath the skin. The overall change in skin temperature is similar even at markedly different environmental temperatures. These effects are accentuated when cooling is increased, such as by having a greater air movement in a climatic chamber. Skin temperature differences become more marked in these conditions and peripheral temperatures, which are initially low, do not change greatly during exercise. However, if sweating occurs during exercise, whatever the environmental temperature, then peripheral blood vessels begin to dilate resulting in a substantial skin temperature rise.

Mean skin temperature is related to environmental temperature whereas body temperature is related to the level of exercise. It appears that the regulating mechanism is controlling deep body temperature during exercise by allowing it to rise to specific levels, depending on the actual heat production. At the same time there is an active control of cutaneous heat loss in relationship to the exercise level. This is a different pattern of regulation from that which is seen when the body is at rest. An explanation of the changes that occur during exercise is still far from clear. Body temperature is normally around 37°C and it has been suggested that there is a mechanism resembling a thermostat in the hypothalamus keeping the level at 37°C. We do not know if there is a central thermostat, and certainly we do not know how it can be changed or reset during exercise itself so as to be related to the immediate level of heat production.

Furthermore, this exercise effect differs from the situation where body temperature is changed by other means. If a subject is immersed in water at 40°C for example, a raised body temperature will result that is perfectly tolerable; indeed the subject may be quite unaware of any increase to his deep temperature although he probably knows that he is flushed and may be sweating. If the subject is immersed in hot water or exposed to high temperature in a climatic chamber there is considerable discomfort by the time a body temperature of 38°C is reached. If this is raised higher to 39°C there is very severe discomfort. In the case of an athlete running in a marathon race with a body temperature of 41 °C, for example, there is no such severe discomfort. We conclude that there are real differences between the effect of a raised body temperature due to environmental conditions such as hot water and one due to exercise.

The effect of immersion in cold water will be described in some detail in Chapter 8. During, and following such immersion, body temperature falls but the surprising thing is that this fall continues on removal from the water. This 'after-drop' of temperature has been explained with reference to the changes that occur to the circulation. Following removal from cold water the intensely constricted skin blood vessels dilate and blood flows through the cooled extremities so that the temperature of the blood returning to the heart is very low. This can have the effect of lowering the deep body temperature but can also have a direct effect upon the heart itself. During the Second World War this decrease in temperature was thought to be responsible for many of the deaths to airmen that occurred a short time after they were rescued from the sea.

It now appears, from results obtained by Golden and Hervey[22] working on the pig, that in these animals the after-drop of temperature occurs even when the heart has been stopped by the injection of potassium chloride when the animal is removed from the cold water. The changes in the deep body temperature in the pig were the same when the circulation was intact and when it had been stopped. These findings show that the after-drop may be due to heat transfer and should therefore be explicable in terms of physics instead of physiology. However, recently Collins *et al.*[23] have shown that in man the after-drop of temperature can be arrested by producing vaso-constriction. It may be concluded that the after-drop is partly due to the physical process of conductive heat exchange and partly due to the cold blood returning from the skin.

Heat loss to the environment

Heat exchange from the skin or clothing surface to the environment is a vital part of the human thermal regulatory process. This energy exchange may take place by conduction, convection, radiation and evaporation. Conduction is the direct exchange of heat from surfaces that are touching, for instance from the soles of the feet to the shoes or from skin or clothes that may be in direct contact with a chair or bed. This mode of heat transfer plays a relatively minor role in overall body heat exchange generally accounting for not more than 1–2 per cent of the total.

Convection is the mechanism of heat exchange whereby the warm skin or clothing exchange heat with the adjacent air causing it to become heated, more buoyant and to rise under the influence of gravity to form a natural convection boundary layer flow. The characteristics of the temperature gradient and thickness of the air layer determine the amount of heat exchange with the surroundings. Convection is markedly increased by body movement or in the presence of wind and this is discussed in detail in Chapter 4.

Electromagnetic waves are generated by molecular vibration from heated surfaces; these waves travel away at the speed of light leaving the surface cooler. If this electromagnetic radiation meets another body it will produce a disturbance to the surface molecules and raise their temperature. This is the basis of radiation heat exchange which is an important route of human body heat loss and described more fully in Chapter 5.

Evaporation is also an important thermoregulatory mechanism and is the method of heat exchange that occurs from exposed liquid where the vapour diffuses away from the surface with the extraction of latent heat from the remaining liquid. Because the human body can regulate sweat output within wide limits, this method of heat exchange can often be the dominating factor in maintaining overall heat balance particularly in strenuous exercise or in hot environmental conditions.

Besides these main routes of heat loss there are several others which play a smaller part. Evaporation from the skin surface can still take place even though there is no active sweating and the skin is not wet. This diffusion of water vapour leads to some heat loss which, together with evaporation of water during expiration, is known as insensible evaporation. Heat may be lost in warming food and air that is taken in, and from the CO_2 that is expired during respiration.

Table 1.6 shows the relative importance of the various modes of heat loss for an adult man averaged over 24 hours. In the absence of strenuous exercise and sensible, or visible, perspiration which leads to a high rate of evaporation, heat exchange occurs mainly by radiation and convection.

Table 1.6 Partition of heat loss by different routes.

Mode of heat loss	Total heat loss (%)	
Insensible water loss		
by breath	11	} 25
by skin	14	
Radiation	37	} 66
Convection	29	
Warming of food and air and liberation of CO_2		9
		100

(Reproduced with permission from *Principles of Human Physiology* by Lovatt Evans[24])

References

1. DuBois, D. and DuBois, E. F. (1915). The measurement of the surface area of man. *Archives of Internal Medicine* **15**, 868–81.
2. Van Graan, C. H. (1969). The determination of body surface area. *South African Medical Journal* **43**, 952–9.
3. Cross, K. W., Gustavson J., Hill, J. R. and Robinson, D. C. (1966). Thermoregulation in an anencephalic infant as inferred from its metabolic rate under hypothermic and normal conditions. *Clinical Science* **31**, 449–60.

4.   Himwich, H. E., Bowman, K. M., Fazekas, J. F. and Goldfarb, W. (1940). Temperature and brain metabolism. *American Journal of the Medical Sciences* **200**, 347–53.
5.   Field, J., Fuhrman, F. A. and Martin, A. W. (1944). Effect of temperature on the oxygen consumption of brain tissues. *Journal of Neurophysiology* **7**, 117–26.
6.   Velten, W. (1880). Ueber Oxydation im Warmbluter bei subnormalen Temperaturen. *Pflügers Archiv* **21**, 361–98.
7.   Krogh, A. (1914). The quantitative relation between temperature and standard metabolism in animals. *Internationale Zeitschrift für Phys. Chem. Biol.* **1**, 491–508.
8.   Bigelow, W. G., Lindsay, W. K., Harrison, R. C., Gordon, R. A. and Greenwood, W. F. (1950). Oxygen transport and utilization in dogs at low body temperatures. *American Journal of Physiology.* **160**, 125–37.
9.   Lynn, R. B., Melrose, D. G., Churchill-Davidson H. C. and McMillan, I. K. R. (1954). Hypothermia: further observations on surface cooling. *Annals of the Royal College of Surgeons of England* **14**, 267–75.
10.   Spurr, G. B., Hutt, B. K. and Horvath, S. M. (1954). Responses of dogs to hypothermia. *American Journal of Physiology* **179**, 139–45.
11.   Deterling, R. A., Nelson, E., Bhonslay, S. and Howland, W. (1955). Study of basic physiologic changes associated with hypothermia. *Archives of Surgery.* **70**, 87–94.
12.   Bickford, A. F. and Mottram, R. F. (1960). Glucose metabolism during induced hypothermia in rabbits. *Clinical Science* **19**, 345–59.
13.   Civalero, L. A., Moreno, J. R. and Senning, A. (1962). Temperature conditions and oxygen consumption during deep hypothermia. *Acta chirurgica Scandinavica* **123**, 179–88.
14.   Hockaday, T. D. R., Cranston, W. I., Cooper, K. E. and Mottram, R. F. (1962). Temperature regulation in chronic hypothermia. *Lancet* **ii**, 428–32.
15.   Bligh, J. (1972). Neuronal models of temperature regulation. In: *Essays on Temperature Regulation* pp. 105–20. Ed by J. Bligh and R. E. Moore, North Holland, Amsterdam and London.
16.   Hensel, H. (1973). Neural processes in thermoregulation. *Physiological Reviews* **53**, 948–1017.
17.   Hensel, H. (1981). *Thermoreception and Temperature Regulation.* Monographs of the Physiological Society No. 38. Academic Press, London.
18.   Stolwijk, J. A. J. and Hardy, J. D. (1966). Temperature regulation in Man. *Pflügers Archiv* **291**, 129–62.
19.   Edholm, O. G., Fox R. H. and Wolff, H. S. (1973). Body temperature during exercise and rest in hot and cold climates. *Archives des Sciences Physiologiques* **27A**, 339–55.
20.   Neilson, M. (1938). Die Regulation der Korpertemperatur bei Muskelarbirt. *Skand. Arch. Physiol.* **79**, 193–230.
21.   Lind, A. R., (1963). Physiological effects of continuous or intermittent work in the heat. *Journal of Applied Physiology* **18**, 57–60.
22.   Golden, F. St. C and Hervey G. R. (1977). The mechanism of the after-drop following immersion hypothermia in pigs. *Journal of Physiology.* **272**, 26–7P.
23.   Collins, K. J., Easton, J. C. and Exton-Smith, A. N. (1982). Body temperature after-drop; a physical and physiological phenomenon. *Journal of Physiology* **328**, 72P.
24.   Lovatt Evans, C. (1952). *Principles of Human Physiology.* J. A. Churchill Ltd.

2

Environmental and Physiological Measuring Techniques

Temperature

Glass bulb thermometers

The liquids used in glass bulb thermometers are generally mercury or alcohol. Mercury is used because it has equal expansion at different temperatures, a low freezing point ($-39°C$), a high boiling point ($356°C$), a high thermal conductivity and a low specific heat; alcohol is also used because, at atmospheric pressure, it does not solidify even at very low temperatures. For these reasons mercury thermometers are used for recording high temperatures and thermometers filled with alcohol for low temperatures. When calibrating such thermometers the melting point of ice is used in preference to the freezing point of water, because distilled water, if undisturbed, may be chilled to a temperature that is several degrees below that at which, if it is not perfectly still, it would freeze. In such circumstances, if suddenly agitated the water will solidify instantly. Also, water which has salt in solution has a freezing point considerably below water which does not contain salt. At the other end of the calibration scale the boiling point of water is a still more variable quantity than the freezing point, and must be qualified with a statement such as 'at mean sea level'.

Glass bulb thermometers are subject to errors which are difficult to eliminate. For instance, the slight plasticity of the glass bulb and the stem require some time for contraction when cooled. Pressure of the mercury column can cause readings to be different when the thermometer is horizontal and vertical, and irregularities in the bore can give rise to variations between thermometers when the size of degrees is not the same.

Mercury thermometers are relatively cheap, simple and compact and by calibration against an accurate thermometer, for instance at the National Physical Laboratory, can lead to an accuracy of $0.01°C$. When greater accuracy is required, or the temperature needs to be measured at an inaccessible site with perhaps a more convenient presentation of the result, other types of thermometer are preferred particularly for application in science and industry.

The resistance thermometer

Temperature rise in a metal is accompanied by an increase in electrical resistance. Resistance can be measured with great precision and this property of metal can form the basis of a very accurate thermometer. Pure platinum is often used and has the advantage that it is chemically unreactive and has a resistance increase of about 40 per cent between 0 and 100°C. A fine platinum wire is coiled around an insulating former and a sensitive ohmmeter measures the changing resistance. Such a thermometer may be calibrated directly in degrees but the resistance change may also be used to define a temperature scale independently of any other thermometer using the following equation:

$$\frac{t}{100} = \frac{R_t - R_o}{R_{100} - R_o}$$

where R is the wire resistance at the temperature indicated by the subscript.

For very precise work, the resistance range can be measured on a Wheatstone bridge network. When the platinum wire is connected to the Wheatstone bridge by copper leads these may also pass near to the heat source and become heated to some extent and thereby introduce an error into the resistance measurement. This may be eliminated by including a pair of dummy leads which experience the same temperature difference, and which are included in the opposite arm of the bridge circuit as shown in Figure 2.1.

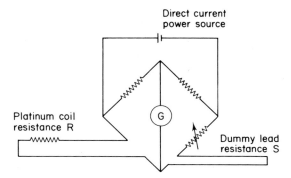

Fig. 2.1 Platinum resistance thermometer circuit. For accurate temperature measurement a fine platinum wire coil, wound on a mica former, forms one arm of a Wheatstone bridge. Dummy leads in the opposite arm of the bridge eliminate errors due to heating of the leads.

Thermocouple thermometers
When two different metals come together at a point of contact an electrical potential difference, depending on the temperature of the junction, is produced. If the circuit is completed with a second junction at a different temperature, electric current will flow around the circuit (Fig. 2.2).

The junction is known as a thermocouple and the thermo-electric effect is now known as the Seebeck effect after the discoverer. The voltage produced across the junctions, or the current flowing in the circuit, may be used to measure temperature. When the cold junction is maintained at a known and constant temperature a calibration graph for the hot junction may be drawn. The graph is not linear so it is necessary to standardize a thermocouple at a sufficient number of points to enable the curve to be drawn.

Thermocouples are very robust, have a low thermal capacity and are very

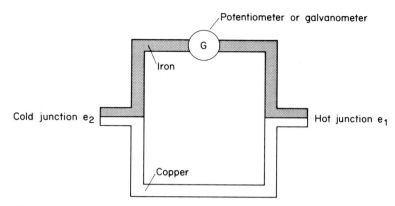

Fig. 2.2 The basis of a thermocouple thermometer. Thermocouples, consisting of two different metals produce electrical potentials dependent on the temperature difference between the two junctions.

compact. Many combinations of metals or alloys can be used depending on the application. For instance platinum-rhodium alloy can be used to measure temperatures up to 1500°C; copper or iron-constantan are other common combinations. Thermocouples lend themselves well to remote temperature sensing and to automated data recording systems and they have been widely used in physiological measurement.

Thermistor thermometers
The thermistor temperature probe is widely used in environmental and physiological temperature measurement, and is particularly suitable for patient-monitoring systems. The thermistor is a semi-conductor which has decreasing resistance with an increase in temperature. A heavy metal oxide (nickel, cobalt and manganese are typical) is compressed to a bead and sintered at a high temperature to form a solid mass. This bead is connected to two leads and encapsulated in a thin glass envelope which can protect it and which is thin enough to allow a rapid thermal response. The thermistor can be connected directly to a suitable bridge circuit or may be built into a transmitter to determine temperature by telemetry. The thermistor resistance falls as temperature rises but in a non-linear fashion. Examples of typical values for resistance are 1400 ohms at 40°C and 2000 ohms at 20°C. These relatively large changes in resistance make the thermistor a very sensitive temperature device.

Infra-red thermography
Thermography is the recording of the temperature distribution of a body from the infra-red radiation emitted by the surface.

The infra-red portion of the spectrum extends from wavelengths at 0.8 μm to 1.0 mm and is sub-divided into areas known as near, middle, far and extreme infra-red (see Fig. 5.2). The electromagnetic energy radiated from the skin surface is proportional to the temperature and has a distribution of wavelengths with the maximum emission occurring in the infra-red region of the spectrum. For technical reasons to do with the detection systems, infra-red thermographic equipment normally used for medical and physiological work detects and measures the radiation in the wavelength band 3–5 μm. The apparatus consists of a special scanner incorporating an infra-red sensitive semi-conductor detector which receives the infra-red radiation; the output from the detector is electronically processed to provide a video picture in either black and white or colour. With a black and white presentation particular temperatures on the skin surface are represented by shades of grey, whilst with the coloured presentation, individual colours represent specific temperatures.

Numerical interpretation of complex thermal images is possible with the lastest thermographic equipment (for example the AGA-Oscar-Pericolor System) in which digital thermographic recording is interfaced with an image analysis computer. Figure 2.3 shows a block diagram of such equipment.

Once thermal images have been digitized and recorded they can be retrieved from the tape store for subsequent numerical computer analysis. The image analysis computer depicted is a hybrid system having a number of inbuilt programs. In addition, extra programming of any function can be carried out by the operator. Much of the software for the programs is based on techniques for earth resource analysis of images from orbiting space satellites.

The following list indicates some of the most commonly used programs suitable for thermographic analysis:

1. In order to calibrate an image, the computer can assign precise temperature levels to specific colours from a given input of the equipment settings, skin emissivity and ambient temperature. Once calibrated, spot temperature measurements may be determined for any part of the thermal image down to 1 pixel size using a cursor under keyboard or 'tracker ball' control.
2. A whole thermal image or any defined portion can be circumscribed and the mean temperature and area evaluated and displayed.
3. Distributions showing the area/temperature frequency are available to be displayed either numerically or graphically on the video display.
4. Sections and profiles showing the variation of temperature along any horizontal or vertical line can be displayed.

If thermograms are to be accurately interpreted a necessary complement to the sensitive infra-red imaging equipment is a well-controlled thermal environment. Minimal air movements with uniformity of temperature ($\pm 0.5°$C) are necessary in the examination room. The room temperature is chosen in relation to the clinical or physiological features being observed. For example, skin blood flow assessment requires temperatures of around 25–27°C whereas 21–23°C is more suitable when identifying subcutaneous thermogenic structures.

Thermography has enabled numerous aspects of skin temperature distribution to be described and visualized; many of these are discussed in Chapter 3 and were either previously unnoticed or particularly difficult to demonstrate. It has also highlighted the caution that must be exercised when using methods

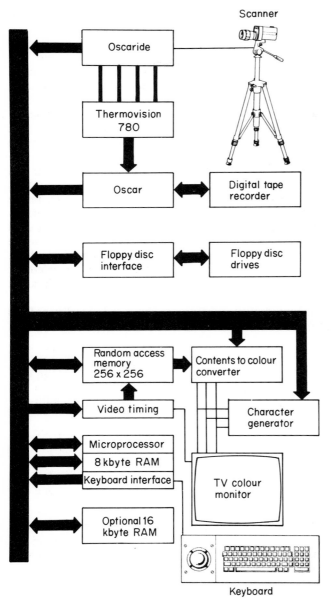

Fig. 2.3 Block diagram of an infra-red thermographic system having a computerized image analysis facility. The upper part of the diagram represents the thermal scanner and electronic digitizing of the image (Oscaride, Oscar and Thermovision 780 blocks) and the lower half of the diagram indicates the components of the image analysis computer. (Redrawn with permission from AGA infrared Systems Ltd.)

such as thermocouple probes to build up temperature distribution patterns over the body surface.

Cholesteric liquid crystals
A relatively new technique for temperature measurement is known as liquid crystal thermography[1, 2]. Liquid crystals are organic substances that can exist in a phase intermediate between a solid and a liquid. In this condition, they may act mechanically as liquids, but exhibit optical characteristics of crystals. The molecules are arranged according to several typical patterns each with differing physical characteristics. One of these patterns or phases is known as cholesteric, and liquid crystals in this phase are suitable for temperature measurement. The term 'cholesteric' is applied to types of liquid crystal that contain mainly compounds of cholesterol.

One of the properties of cholesteric liquid crystal is known as circular dichroism, so named because of the arrangement of the molecules. This means that if a beam of unpolarized white light encounters a layer of liquid crystal it is broken into two components; one of these is transmitted and the other reflected. If the wavelength of the reflected beam is in the spectrum of visible light, the surface of the liquid crystal appears coloured. Another property, and one which enables these crystals to be used in thermal studies, is the influence that temperature has on the wavelength of the reflected light, and therefore on its colour. Liquid crystal thermography is the recording of the temperature distribution over hot surfaces by covering them with the crystals in one of several forms and observing the colour changes that occur. They may be used undissolved or in solutions and may be applied to a surface over a matt black paint base. Alternatively, the crystals can be in the form of micro capsules sandwiched between polyester sheets and in this form they are often used for clinical contact thermography. In skilled hands the technique can help in the diagnosis of pathological conditions, for instance breast cancer; other uses have included disposable oral thermometers and 'vein finders' where, if placed on the skin surface, the track of subcutaneous veins is outlined.

The globe thermometer
This device is used to measure the mean radiation from the surroundings and consists of a 15 cm diameter metal sphere painted matt black and containing a mercury in glass thermometer with its bulb positioned at the centre of the sphere (Fig. 2.4). The temperature that the instrument records will depend on the environment in which it is placed. If the surrounding walls and surfaces are warmer than the air, the temperature recorded by the thermometer inside the globe will be above the air temperature; if the surroundings are cooler than the air, the globe thermometer temperature will be below that of the air. Globe temperature may also be influenced by cooling due to air movement. The globe thermometer will equilibrate with the environment after about 20 minutes exposure.

Measurement of globe thermometer reading, air temperature and air velocity can be used to define the mean radiant temperature of the environment. This is the uniform temperature at which a black surface would radiate

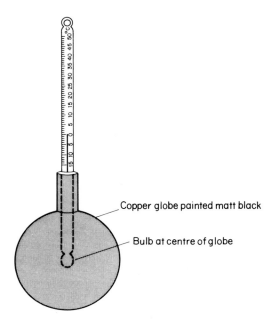

Copper globe painted matt black

Bulb at centre of globe

Fig. 2.4 The globe thermometer consisting of a mercury in glass thermometer whose bulb is at the centre of a 15 cm diameter blackened copper globe.

with an intensity equal to the mean observed. Bedford and Warner[3] computed mean radiant temperatures from globe thermometer, air temperature and air speed measurements and these calculations are now embodied in nomograms.

A globe thermometer reading itself is inadequate as an index of the thermal environment which can be illustrated as follows. When the air temperature is the same as the surrounding surfaces the globe thermometer will read air temperature irrespective of air velocity over the globe; if the surroundings are cooler than the air the globe thermometer will give a lower reading than air temperature. However, if the air temperature and radiation remain constant, the globe reading will approach that of the air as air velocity increases. If the human body were subjected to these two environments the one causing the greatest cooling would be the one registering the highest temperature on the globe thermometer.

Humidity

The continual process of evaporation means that the atmosphere always contains water vapour. The quantity of vapour is referred to as the humidity and there is a limited amount that the air can hold; once this limit has been reached the air is said to be saturated. The mechanism by which this happens can be illustrated by considering a closed jar that is partially full of water. The molecules in the liquid are in continual random motion and there are frequent collisions with other molecules and with the free surface of the water in the jar. Some of the fast-moving molecules may collide with the exposed water surface and break free to enter the air space above as gaseous molecules. As the more energetic molecules leave the water they take energy with them which results in a fall in the average temperature of the liquid. This process forms the basis for the concept of latent heat, which is defined as the heat required to convert unit mass of a substance from one state to another at a constant temperature. Substances may have a latent heat of melting (or fusion) and a latent heat of evaporation. For water, the latent heat of fusion is 335 J/g and the latent heat of evaporation is 2394 J/g at 45°C rising to 2513 J/g at 5°C. If the area above the liquid is not contained, the escaping molecules may move out of the container and diffuse into the environment and evaporation will continue until there is no liquid left. An exposed liquid is usually a few degrees cooler than the surroundings due to the removal of latent heat by the evaporative process. The rate of evaporation is temperature dependent because at higher temperatures the molecules have more energy and the velocity of their escape from the free surface is higher.

If the space above the liquid is closed the molecules of water vapour can no longer escape and they build up above the liquid.

From time to time, moving molecules in this space penetrate the free surface and return to the liquid, and as more and more enter the space above the liquid the numbers returning will also increase until a state is reached when the number of molecules evaporating is equal to the number condensing back into the liquid. When this occurs, the space above the liquid is said to be saturated. The molecules constituting the vapour produce a pressure which increases as the density of the vapour becomes larger. The pressure exerted when the vapour is saturated is steady and is the saturated vapour pressure or SVP.

The humidity of any given environment may be expressed as the partial pressure of the water vapour in the air. If this partial pressure is divided by the saturation vapour pressure, at the same temperature as the air, the quotient is known as the relative humidity (RH). It is conventional to express this ratio as a percentage with 100 per cent RH corresponding to saturation.

Relative humidity is particularly important in physiological studies because moisture exchange with the environment, particularly when considering evaporative heat loss and sweating, is dependent on the relative humidity rather than the absolute humidity (which is the mass of water vapour contained per unit volume of air).

Another method of expressing the water vapour content of the atmosphere is to define the dew point. This is the temperature at which dew begins to form when air is slowly cooled. At this temperature, the water vapour present in the air is just sufficient to saturate it and any further cooling gives rise to condensation. This effect is the basis for a very accurate estimation of humidity by the use of a dew point hygrometer. Figure 2.5 illustrates the basic principles of such a hygrometer in which a container is cooled by being filled with ether which is made to evaporate by a controlled air stream. The outer surface of the container is highly polished so that the moment of condensation

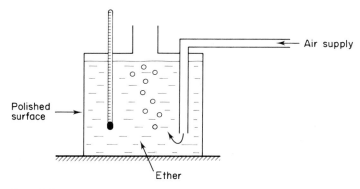

Fig. 2.5 Determination of dew-point by condensation of water vapour on a cooling, polished surface.

can be easily seen. At this point. the temperature of the evaporating ether directly measures the dew point.

Wet bulb temperature
As already mentioned, an exposed liquid is usually a few degrees cooler than the surroundings because of the extraction of latent heat by evaporation. After an initial drop in temperature, heat flows into the liquid from the environment at the same rate as it is lost by evaporation and, as no other heat source is supplied, the air temperature will fall. After some time, the air will become saturated by the vapour from the evaporating water because the temperature is falling. This principle is used in the wet bulb thermometer which has the bulb itself covered with a soft cloth soaked in clean water. As moisture evaporates from the wet cloth it takes heat from the bulb of the thermometer which in turn is extracted from the atmosphere; the mercury in the thermometer gives a lower reading than dry bulb temperature according to the rate of evaporation. In order not to have an artificially high reading by incomplete evaporation, the wet bulb should be in a moving airstream.

Hygrometers
The difference between wet and dry bulb temperatures is used in a hygrometer to give a measure of atmospheric humidity. The psychrometer or sling hygrometer is perhaps the most convenient instrument for humidity measurement employing this principle. There is no exact theory to describe the processes involved and the method is an empirical one. This device is shown in Figure 2.6, where two similar mercury thermometers are mounted in a sling that can be rotated. One bulb is covered with a wetted fabric so that when the sling is rotated the moving air can cause rapid evaporation from the thermometer globe which will therefore give a lower reading than the adjacent dry bulb thermometer subjected to the same airstreams. The difference in temperature between the two thermometers correlates with the degree of atmospheric humidity and may be determined from empirical tables. Each particular configuration of wet and dry bulb hygrometer has its own characteristics from which to determine the actual humidity.

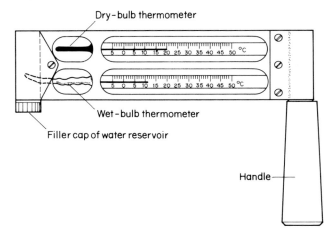

Dry-bulb thermometer

Wet-bulb thermometer

Filler cap of water reservoir

Handle

Fig. 2.6 The sling hygrometer for measuring humidity consists of two thermometers that are turned by vigorously swinging the handle and exposing the thermometers to rapid air movement. The bulb of one thermometer is covered with a silk sleeve that is kept moist and this will record wet bulb temperature. The atmospheric humidity is determined by special tables from the difference in reading between the two thermometers.

Another psychrometer in which the wet and dry bulb thermometers are enclosed in a metal casing and where a small motor operated fan draws air past the thermometers is known as the Assmann hygrometer. Figure 2.7 shows such a psychrometer where the thermometer bulbs are protected from radiation by cylindrical shields.

Hair hygrometer
Human hair has the property that its length is dependent on atmospheric humidity. This forms the basis for the hair hygrometer, which consists of one or more strands of hair and a mechanism whereby any change in length of the strands, due to changes in humidity, can cause an indicator to move across a dial. Such a device requires initial calibration and frequent checking or setting thereafter. They are not recommended for precise work but in environments with slight changes in humidity they make useful indicators.

Corona-discharge hygrometer
A number of other techniques are available for measuring humidity in the atmosphere, many of them relying on the absorption of water vapour by hygroscopic materials. Most of these methods have a slow response time which often limits their usefulness. One way to overcome such a disadvantage was developed at the Royal Aircraft Establishment[4] and is based upon the corona discharge characteristics which are sensitive to the water vapour content of the air. The principle of this device is illustrated in Figure 2.8 in which a thin wire, acting as an anode, and raised to a potential of some 4 kV, is placed centrally in a cylindrical cathode. If this cathode is insulated from

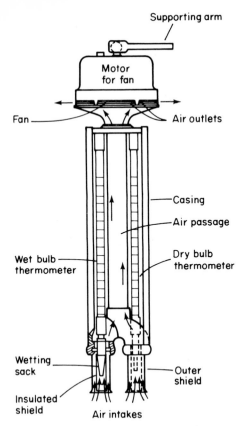

Figure 2.7 The Assmann Hygrometer. Humidity is determined from the readings of wet and dry bulb thermometers which are exposed to an airstream produced by a clockwork or electrical fan.

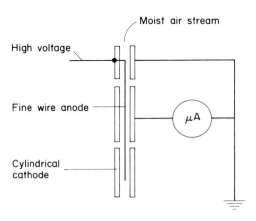

Fig. 2.8 Diagram to show the principle of humidity measurement by means of a corona discharge. The current passing between a fine wire anode maintained at high voltage and a cylindrical cathode tube is proportional to the water content of the air that is drawn through the cathode.

other parts of the apparatus and air is passed down the cylinder, the resulting corona discharge gives rise to a small current between the cathode and the earth. This device can form a simple and inexpensive humidity meter with the benefit of a rapid response time.

Air movement

The pitot-static tube

Perhaps the most accurate method for measuring air velocity, particularly where the airspeed is quite high (in air-conditioning ducts, for instance) is by means of the pitot-static tube. This consists of two tubes, one inside the other, which are bent at right angles as shown in Figure 2.9. The outer tube has a series of circumferential holes, known as static pressure holes, some way back from the nose piece. The open end of the central tube faces directly into the airflow. Outlets for both the central and outer tubes are taken independently to either side of a manometer. The device makes use of the Bernoulli equation which states that for any part of the air stream $P + \frac{1}{2}\rho V^2 = $ constant where P, ρ and V are the air pressure, density and velocity, at any point.

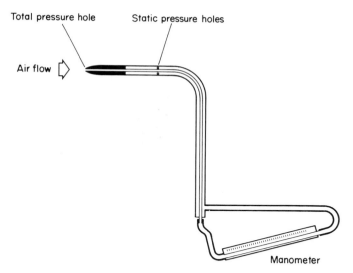

Fig. 2.9   The pitot-static tube for measuring air velocity comprises a double tube bent at right angles and facing into the airstream. The inner tube facing the flow is subject to the impact of air velocity and communicates total pressure whilst the outer tube, with holes around the circumference, measures static pressure. The difference between these pressures, measured with the manometer, is a function of air velocity.

If this equation is applied at two places in the flow, the difference between the pressure at the front of the nose piece (the so-called 'velocity' pressure) and the static pressure (measured at the circumferential holes behind the nose piece) can be directly related to the air velocity. The final equation is of the form:

$$P_1 - P_2 = \tfrac{1}{2}\rho V^2$$

where $P_1 - P_2$ is the measured pressure difference and consequently

$$V = \sqrt{\frac{2}{\rho}(P_1 - P_2)}$$

with V being a function of the pressure difference and the air density only.

The following sections describe anemometers which are suitable for measuring airspeed in situations where the pitot-static tube would be impracticable or insensitive. However, devices such as the vane, cup and hot wire anemometers require calibration from time to time and this should ideally be carried out using a pitot-static tube in a wind tunnel.

The cup anemometer
For the measurement of mean wind speed out of doors, a cup anemometer is generally used (Fig. 2.10). This consists of arms, attached to hollow cups

Fig. 2.10 The cup anemometer suitable for wind speed measurement out of doors.

which rotate horizontally about a vertical axis. The speed of rotation is measured either electrically or mechanically with the resultant wind speed being displayed on a calibrated dial.

It is useful to be able to relate descriptions of outside wind to the actual air velocity and one of the earliest attempts at this was made by Admiral Beaufort who, in 1806, devised a scale having a relation to the pressure of wind upon the sails of a ship to the amount of canvas which a ship should carry. Table 2.1 shows this scale which is divided into 12 speed ranges. Alongside each description of wind force is given a description of effects that may characterize the Beaufort scale on land.

Table 2.1   The Beaufort scale of wind force.

Wind speed (mph)	Beaufort scale	Wind force	Land description
1	0	Calm	Smoke rises vertically
1–3	1	Light air	Wind direction shown by smoke and not by wind vane
4–7	2	Slight breeze	Wind felt on face, leaves rustle, ordinary vane moved
8–12	3	Gentle breeze	Leaves and small twigs in constant motion, wind extends high flag
13–18	4	Moderate breeze	Raises dust and loose paper, small branches are moved, snow begins to drift
19–24	5	Fresh breeze	Small trees in leaf begin to sway, created wavelets form on inland waters
25–31	6	Strong breeze	Large branches in motion, whistling heard in telegraph wires, high drift occurs
32–38	7	High wind	Whole trees in motion, inconvenience felt when walking against wind, visibility obscured by drifting snow.
39–46	8	Gale	Breaks twigs off trees, generally slows progress
47–54	9	Strong gale	Slight structural damage occurs— chimney pots and slates fall off roofs
55–63	10	Whole gale	Inland trees uprooted
64–72	11	Storm	Widespread damage
73–82	12	Hurricane	Hurricane

The vane anemometer

This device is suitable for measuring air velocities that may be encountered in ventilation ducts or grills and consists of a light vane wheel which is accurately balanced to move easily on its bearings. In a mechanical vane anemometer the number of rotations of the vane in a given time is displayed on a counter.

A simple calculation then gives the mean air velocity. A more convenient form of this anemometer senses the speed of the rotating vanes electronically and the air velocity is directly displayed on a meter (Fig. 2.11).

Fig. 2.11   A vane anemometer with the vane speed electronically sensed to produce a direct reading of airspeed.

Thermal anemometers

If a sensing element is heated electrically and then exposed to an airstream, the temperature difference between the element and the air can be calibrated to produce a measure of air velocity.

In the heated thermocouple anemometer the differential voltage between a heated and unheated thermal junction exposed to an airstream can produce an accurate measure of air velocity. Another device is known as a hot wire anemometer where a very fine wire is heated to a high temperature; the power required to maintain this temperature constant when the wire is cooled by an airstream gives an accurate measure of the air velocity. This type of anemometer can have an extremely short response time and can be used in very detailed studies of turbulent airstreams. With suitable damping, it is useful for mean airspeed measurements and has been employed successfully at very low flow rates when measuring velocity profiles in the human natural convection boundary layer flow[5].

The Kata thermometer

The Kata thermometer (Fig. 2.12) was invented by Leonard Hill[6, 7] many years ago but it is still a very useful device for environmental investigations. It can be used as an anemometer to measure air movements in ventilated areas, particularly when the air is slow moving, and has also been used as a comfort meter to assess the cooling power of the environment. The thermo-

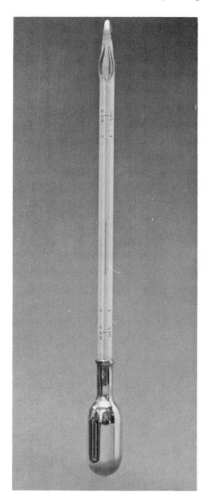

Fig. 2.12 The Kata thermometer is used to measure slow air movements in ventilated areas. It consists of a large globe, filled with coloured spirit. The stem is marked at points corresponding to bulb temperatures of 100°F (38°C) and 95°F (35°C). When the bulb is heated and then allowed to cool the time taken for the spirit to fall between the two marks is a measure of the cooling power of the environment and air movement may be estimated from special charts.

meter has a large bulb filled with coloured spirit; it is used by immersing the bulb in hot water until the alcohol has expanded to fill the reservoir at the top of the stem. The bulb is then removed from the hot water and dried and the thermometer is suspended in the air. The time taken for the liquid to fall between two marks on the stem equivalent to 100°F (38°C) and 95°F (35°C) respectively is then used as a measure of the rate of cooling of the thermometer. When used as a dry bulb instrument the cooling rate is affected by radiation and convection from the heated globe. It may also be used as a wet bulb device, in which case a wet silk sleeve net is fixed tightly around the bulb. When used in this way, the bulb loses heat by evaporation as well as by convection and radiation. Air movement past the globe increases the cooling power of the Kata thermometer and the wet and dry cooling powers may be used as an index of the cooling effect of the environment on the human body.

When the cooling power is used as an index of air movement a number of variations are available. For measuring airspeeds in hot environments a high temperature version (130–125°F (54–52°C)) is used. If an environment includes radiation from surfaces much hotter or colder than the air, Kata thermometers with silvered bulbs may minimize errors due to radiation. Each instrument has a calibration factor which, when divided by the average time of cooling in seconds, results in a cooling power expressed in millicalories/cm²sec. Air velocity from the dry Kata reading may be determined from a knowledge of the cooling power and environmental temperature using empirical formulae or specially constructed charts.

Body temperature measurement in man

Oral temperature

The most common method of measuring body temperature is with a mercury and glass thermometer placed under the tongue. This technique is perfectly valid but there are a number of factors which have to be taken into account to ensure a reasonable accuracy in the measurement. Due regard must be paid to the condition of the surface temperature in the mouth which may be affected by breathing or speaking, by a hot or cold drink or by the ingestion of very hot or cold food such as ice cream. In all of these cases the surface temperature inside the mouth can remain markedly different from body temperature for periods as long as 20–30 minutes and therefore the oral temperature can sometimes be very misleading. Other precautions are concerned with the thermometer itself, which very often takes a considerable time to reach equilibrium. If such a thermometer is read in two minutes and then at, say 30 second intervals thereafter, it may take five minutes before the body temperature reading is constant. In practice such precautions are seldom observed, and so when oral temperature readings are reported there should be some caution in interpreting them with any precision. Instead of using a mercury and glass thermometer for measuring temperature in the mouth it is possible to use either thermistor or thermocouple probes.

Rectal temperature

The measurement of deep body temperature in the rectum is used clinically and experimentally and has been considered to be more reliable than mouth temperature. This is because the local conditions in the rectum do not affect the measurement of temperature as much as local conditions in the mouth. However, position does matter in the rectum, as was shown many years ago by Mead and Bonmarito[8] who found that the location of the temperature probe affected the reading although by not more than 0.1–0.2°C. A problem of rectal temperature measurement is that either a thermocouple or thermistor probe must be inserted to a considerable depth, of the order of 9 cm, and many people find this objectionable. A mercury and glass thermometer may be used, but considerably more care has to be taken to ensure against breakage. This latter method is still used clinically in many places, but at least 3 minutes must be allowed for the temperature to come to equilibrium. The actual temperature measured in the rectum under good conditions is in general about 0.5°C higher than mouth temperature. This will be referred to again when other methods of body temperature measurement are compared for accuracy.

Aural temperature
The ear or aural site may also be used to give a measure of body temperature. A thermistor or thermocouple is inserted into the external auditory meatus and down into the auditory canal to the drum, the tympanum. Touching the drum is painful and when this happens the thermocouple bead or thermistor is slightly withdrawn. The problem is to ensure that the temperature measured is representative of deep body temperature. It is important to provide insulation of the ear in moderately cool and certainly in cold conditions. This insulation has to be at least 1–2 cm thick otherwise there is a gradient of temperature along the auditory canal and the deep body temperature may not be reached at the point of the tympanum. Even in warm conditions it is wise to have some insulation over the ear itself.

The radio pill
Other methods of deep body temperature measurement include the use of the radio pill, designed by Wolff[9]. This consists of a very small radio transmitter made into a temperature sensitive capsule which can be swallowed quite easily. The transmitted signal carrying the temperature information is picked up with a receiver placed close to the body surface. The pill can record temperature with an accuracy which may be as good as ± 0.1°C and is useful because measurements can be made over very long periods of time without disturbing the subject. It is effective once it has left the stomach and departed on its journey down the alimentary tract, first to the duodenum and then into the small intestine. There does not appear to be any difference in the temperature recorded in any part of the alimentary canal until the rectum is reached. This journey takes about 24 hours on average and the battery can last between 24 and 36 hours; signals can therefore be received at least over the period of one day. The pill has proved of great value in field work but has the disadvantage that it is expensive and steps therefore have to be taken to recover it; this can be done quite sucessfully but requires special arrangements!

Oesophageal temperature
Apart from measurements in the intestinal tract using the radio pill, body temperature has been measured in the oesophagus or in the stomach using thermocouples or thermistors inside narrow bore rubber tubing; such catheters can be passed without any great difficulty and can remain *in situ* for periods of some hours. In certain cases these probes can be extremely useful but they do require more tolerance on the part of the subject than a radio pill of which a subject is totally unaware once swallowed.

Urine temperature
The temperature of freshly voided urine is the same as that of the urine in the

bladder and is very close to body temperature. Fox has designed a urinometer[10] where urine is passed over the surface of a thermometer which is read straight away. This is a valuable method of measuring body temperature, especially in field-work, where the subject has to void urine in any case.

Transcutaneous deep body temperature

A further method, also designed by Fox *et al.*[11] uses a skin temperature measured when there is zero heat transfer between the area of skin at which the measurement is taken and the deep tissues. In this technique a multilayer flexible pad is used which contains a thin electrical heating element and two thermistors embedded at different depths in the pad. This sensor is about 6 cm square and 0.6 cm thick and covered with a coating of silicone rubber (Fig. 2.13(a) and (b)). It is attached to the skin surface, for instance at the

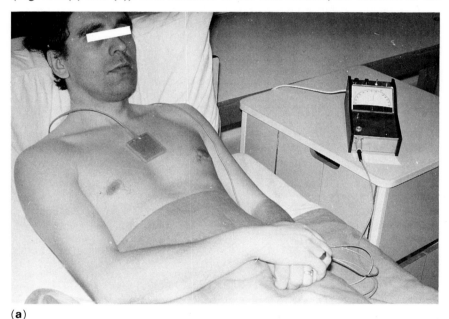

(a)

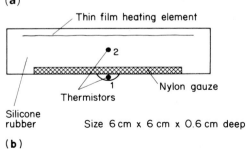

(b)

Fig. 2.13 Transcutaneous temperature measuring probe attached to a subject (a) and in schematic form (b).

lower chest, and a control circuit enables power to be supplied to the heating element. This heating continues until there is equilibrium between the temperatures measured by the two thermistors embedded in the probe. At this stage there is no heat transfer from the area of the skin covered by the sensor which becomes a region of zero heat flow from the core across the body shell. Deep body temperature is consequently 'brought' to the skin surface and recorded as the equilibrium pad temperature. This method produces generally slightly lower values than rectal temperature but the response to changes in deep body temperature can be rapid and the results are relatively independent of the environment.

Comparison of body temperature measurements
It is important to try to compare all of these various methods of temperature measurement and this is done in Figure 2.14. As can be seen, mouth and ear temperatures are in general rather close together. Rectal values are higher than both, and the radio pill is usually closer to rectal than either mouth or ear. Urine temperature is usually close to that of the ear and the mouth and is always lower than rectal temperature. Oesophageal and gastric temperatures are very close to those measured with the pill. When body temperature changes there is a more rapid response in ear and mouth than in the rectum, which may lag behind for up to 10–15 minutes. Pill and urine temperatures

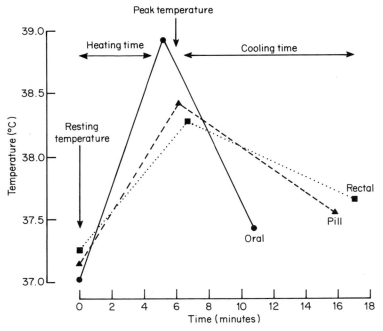

Fig. 2.14 Body temperature measured in the mouth, the rectum and using the radio pill. The divergences are shown during a period of body heating and cooling.

are intermediate but urine is perhaps closer to the pill measurement than to the ear. This also means that if body temperature rises and then falls, the mouth and ear temperatures rise higher than either the rectal, the pill or the urine temperatures. It is a matter of some debate as to which of these methods most closely records deep body temperature. In general it may be fair to say that they all reflect a different body temperature and it is really for the investigator to decide which fulfils his purpose most conveniently. Figure 2.15 illustrates several body temperature measuring devices.

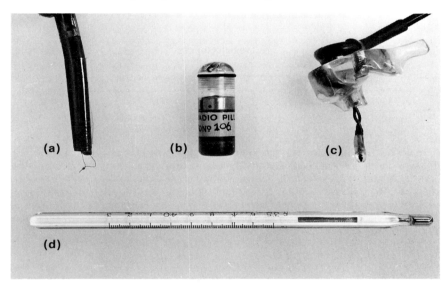

Fig. 2.15 A comparison of temperature measuring devices. A skin thermister (a), a radio pill (b), an aural thermistor probe (c) and a clinical thermometer (d).

'Normal' body temperature
Body temperature is mentioned and discussed throughout this book and only a brief review of the level of body temperature and its usual variation is given here. It is conventionally stated that normal temperature is 37°C (98.4°F) and mercury-in-glass clinical thermometers are appropriately marked. This strictly only applies to mouth temperature; the levels to be found at other sites are not necessarily the same. Rectal temperature is on average about 0.3–0.5°C higher; urine temperature is intermediate between mouth and rectum. Aural temperature is close to or slightly higher than mouth temperature. There are some regional differences in body temperature depending upon levels of heat production but these are small owing to the efficiency of the blood circulation in distributing heat and smoothing out the differing levels of temperature.

The level of 37°C is characteristic of mouth temperature, and is the mean figure obtained from measurements of many individuals. There is a normal range of approximately ±1°C. Furthermore, the level is affected by many factors as described in detail throughout this book. These factors include the degree of muscular activity (body temperatures of 40°C and over can occur in marathon running), exposure to hot or cold environments, the ingestion of hot or cold food or drink, immersion in hot or cold water and exposure to differing atmospheres specifically the effect of helium. All these factors are considered in subsequent chapters and Figure 2.16 illustrates body temperature variation.

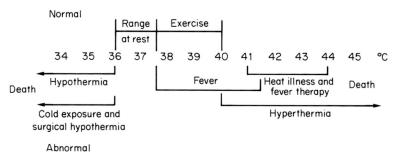

Fig. 2.16   Range of body temperature. The usual range is from approximately 36 to 40°C; at temperatures above 41°C there is an increasing risk of death. 'Heat Illness' includes the effects of high environmental temperature, when the body may be heated above 41°C, and can lead to heat stroke. 'Fever' is usually due to an infection. The body can be cooled to very low temperatures (below 20°C) and still recover. This is done for certain operations (e.g. surgical hypothermia for heart and brain operations).

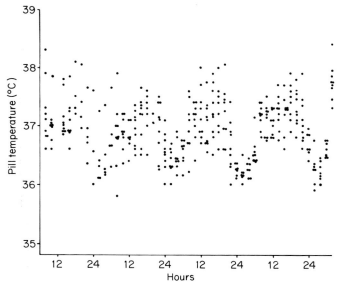

Fig. 2.17   Body temperature, measured with the radio pill, in nine subjects during a period of four days. The circadian rhythm of the temperature is evident. At night, body temperature can be as low as 36°C or below. (Reproduced with permission from Edholm, O.G., Fox, R.H. and Woolff, H.S. (1973). Body temperature during rest in cold and hot climates. *Archives des Sciences Physiologiques* **27**, 3.)

There is a further source of variation and that is the time of day. There is a daily variation in the level of body temperature, now generally described as the circadian rhythm (from *circum diem*). The amplitude varies between individuals ranging from 0.7 to 1.5°C. Body temperature can be as low as 35.6°C in some subjects during the night (asleep) and as high as 37.6°C in others, even at rest. This rhythm persists in a fasting state and also in subjects confined to bed. When the life pattern is changed from night time sleep and daylight waking to the reverse, with activity at night and sleep during the day, the original circadian rhythm initially persists. Eventually the rhythm will adapt to the new life style. It can be concluded that the rhythm is endogenous; the fundamental mechanism is still not clear. There is evidence that there is an associated change in the pattern of heat loss, but this would only partly account for the size of the variation[12]. Changes in the level of the temperature set point and other possible mechanisms are discussed by Minors and Waterhouse[13]. Figure 2.17 shows an example of the rhythmic temperature changes in a group of soldiers measured 4-hourly.

Heat transfer measurement

The Hatfield heatflow meter

Local heat transfer from the skin or clothing surface is an important measurement in determining the body's overall thermal response to an environment. The Hatfield heatflow meter[14] was able to measure directly the passage of heat energy when attached to the skin or clothing. In its original form it consisted of a thin disc (12 mm diameter and 1.5 mm thick) of a tellurium alloy having a fine copper gauze attached to both of its sides. The copper gauzes were connected by fine copper wires to a mirror galvanometer. When the disc was attached to a hot surface heat flow through the disc created a temperature difference which was proportional to the heat flow and the thickness of the disc and inversely proportional to the thermal conductivity. The thermo-electric current set up in this way gave a direct measure of the heat transfer through the disc. The tellurium alloy produced a much greater thermo-electric effect than more usual materials and a heat flow of $10 \, W/m^2$ could be recorded at a voltage of $6.0 \, \mu V$ on the galvanometer. In its commercial form the disc was supplied from the manufacturers with individual calibration.

One of the first uses of the Hatfield disc was to measure the thermal conductivity of human fat and muscle. Hatfield and Pugh[15] obtained specimens of human tissue from post-mortem rooms and in one series of tests they found that the conductivity of human muscle varied considerably but that fat had a more constant value. Typical mean values for the conductivity of fat were $0.2043 \, W/m.°C$ and $0.3852 \, W/m.°C$ for muscle indicating that fat was a better thermal insulator than muscle.

In life, skin temperature and heat transfer to the environment are dependent on such tissue conductivities which will be modified by the presence of muscle and of skin blood flow. These factors contribute towards the temperature patterns revealed by infra-red thermography and discussed in Chapter 3.

There is considerable variability in individual skin temperature patterns and it is possible that variations in tissue conductivity may, to a large extent, account for this.

Surface plate calorimeter

When measuring local heat transfer coefficients directly from the body surface, it is necessary to have an instrument sensitive enough to measure losses by convection and radiation in the human natural convection boundary layer flow. Toy[16] considered a number of alternative measurement techniques for this use and critically examined the Hatfield heatflow disc. He concluded that the Hatfield disc could give some measure of local convection heat losses and could be comfortably attached to the human body but that when measuring free convection from a heated horizontal surface the disc might over-estimate heat loss by as much as 25 per cent; in forced convection it could under-estimate heat transfer by a variable amount. In addition, the thickness of the disc was often a large percentage of the boundary layer thickness over the body and consequently disturbed the flow patterns and thereby the convective heat transfer. For these reasons, the Hatfield disc was not considered sufficiently accurate to give a confident measure of convective heat transfer from the human microclimate and a new instrument was designed specifically for this purpose. This is known as a surface plate calorimeter[17] and is shown in Figure 2.18. A straingauge used as an electric heating element is attached to one surface of a copper plate 0.05 mm thick and approximately 8×12 mm in area. The principle of the calorimeter is that once attached, it replaces that part of the skin that it covers. Electric power applied to the heating element raises the temperature of the copper plate to that of the skin surface; once equilibrium has been reached, the power required to maintain the plate at skin temperature is a measure of the heat loss by convection and radiation from the skin covered by the calorimeter.

In its original form the calorimeter was manually operated and was used to measure heat transfer coefficients from the body in a number of situations which are described throughout this book. Recently, the technique has been refined so that the electric power supplied to the heating element is precisely controlled from the output of a differential amplifier which compares the outputs from the thermocouple junctions on the skin and within the calorimeter[18]. In this way, the control system can automatically follow any changes of temperature and heat flow that may occur over the body surface, and the local rate of heat transfer is displayed directly in digital form.

Partitional calorimetry

A method for measuring the heat loss from the human body in terms of convection, radiation and evaporation was developed by Winslow, Herrington and Gagge in 1936[19]. The method was called 'partitional calorimetry' and involved the use of a booth consisting of nine panels of copper, with accurate control of air temperature and humidity. Heating was effected by infra-red radiant heat; the copper walls reflected 98 per cent of the radiation and remained cool. Hence, wall and air temperature could be held constant

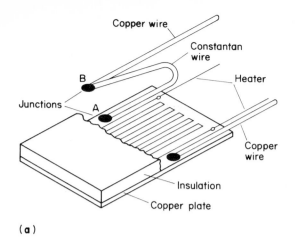

(a)

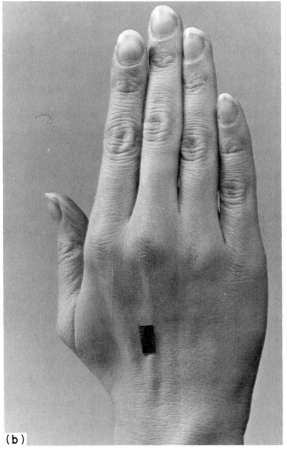

(b)

Fig. 2.18 Surface plate calorimeter in diagrammatic form (a) and shown attached to the hand (b). (Reproduced with permission from Toy, N. and Cox, R.N. (1974). *Revue Générale de Thermique*.)

or varied independently of each other, so heat loss by radiation could be separated from that lost by convection.

The great contribution by Winslow *et al.* was they succeeded in making all the necessary environmental and physiological measurements which included oxygen consumption, ambient temperature, humidity, mean skin temperature, body temperature and weight change during the experimental period[20]. In addition, globe temperature, clothing surface temperature and clothing insulation need to be included.

The original method is now of historical interest only as other and more convenient techniques have been introduced. As described in this book, the use of thermography has made it possible to measure radiation heat exchange and convection heat loss can be assessed using heat flow gauges and fast responding calorimeters and the Schlieren technique.

References

1.   Fergason, J.L. (1964). Liquid crystals. *Scientific American* **211,** 77–85.
2.   Crissey, C.T., Gordy, E., Fergason, J.L. and Lyman, R.B. (1964).—A new technique for the demonstration of skin temperature patterns. *Journal of Investigative Dermatology* **43,** 89–91.
3.   Bedford, T., & Warner, C.G. (1934). The Globe Thermometer in studies of heating and ventilation. *Journal of Hygiene*, Cambridge **34,** 458–73.
4.   Maskell, B.R. (1970). The effect of humidity on a Corona discharge. *Royal Aircraft Establishment*. Technical Report 70106.
5.   Clark, R.P. (1973). *The role of the human micro-environment in heat transfer and particle transport*. PhD Thesis. The City University, London.
6.   Hill, L. and Vernon, H.M. (1923). *The Kata thermometer in studies of body heat and efficiency*. Medical Research Council Special Report Series No. 73.
7.   Hill, L., Griffith, O.W. and Flack, M. (1916). The measurement of the rate of heat loss of body temperature by convection, radiation and evaporation. *Philosophical Transactions of the Royal Society of London, Series B* **207,** 183.
8.   Mead, J and Bonmarito, C.L. (1949). Reliability of rectal temperature as an index of internal body temperature. *Journal of Applied Physiology* **2,** 97–109.
9.   Wolff, H.S. (1961). The radio pill. *New Scientist* **12,** 419.
10.   Fox, R.H., Macdonald, I.C. and Woodward, P. (1973). A hypothermic survey kit. *Journal of Physiology* **231,** 4–6P.
11.   Fox, R.H., Solman, A.J., Isaacs, R., Fry, A.J., and, Macdonald, I.C. (1973). A new method for monitoring deep body temperature from the skin surface. *Clinical Science* **44,** 81–86.
12.   Aschoff, J. and Heise, A., (1972). Thermal conductance in man, its dependence on time of day and on ambient temperature. In: *Advances in Climatic Physiology* pp 334–48. Ed by S. Itoh, K. Ogata and H. Yoshimura. Igaku Shoin, Tokyo.
13.   Minors, D.S. and Waterhouse, J.M. (1981). In: *Circadian rhythms and the human*. John Wright and Sons, Bristol.
14.   Hatfield, H.S. (1953). An apparatus for the measurement of the thermal conductivity of biological tissue. *Journal of Scientific Instruments* **30,** 460–1.
15.   Hatfield, H.S. and Pugh, L.G.C.E. (1951). Thermal conductivity of human fat and muscle. *Nature* **168,** 918.
16.   Toy, N. (1976). *Local free and forced convection heat transfer from the human body*. PhD Thesis. The City University, London.

17.   Clark, R.P., Cox, R.N. and Toy, N. (1972). A surface plate calorimeter for measuring local heat transfer in free and forced convection within the human micro-environment. *Journal of Physiology* **223,** 10–12P.

18.   Barnett, T.G., Clark, R.P. and Stothers, J.K. (1981). An electronic control system applied to surface plate calorimetry. *Journal of Physiology* **313,** 2P.

19.   Winslow, C-E.A., Herrington, L.P. and Gagge, A.P. (1936). A new method of partitional calorimetry. *American Journal of Physiology* **116,** 641–52.

20.   Gagge, A.P. (1972). Partitional calorimetry. In: *Physiological Adaptations.* Ed by M.F. Yousef, S.M. Horvath and R.W. Bullard. Academic Press, London and New York.

3

Skin Temperature

As a preliminary to the study of the way in which the body can lose heat to the environment, it is necessary to investigate the temperature found on the skin or clothing surface. It is the difference between this temperature and that of the surroundings that enables non-evaporative heat exchange to take place. In most situations the skin is warmer than the surroundings and the direction of heat flow is away from the body; however in some conditions, for instance in hot climates, this may be reversed and the body may gain heat if the air is hotter than the skin. In such circumstances the body will still be able to lose heat by sweating and can achieve thermal balance, but at a higher deep body temperature.

The classic description of the distribution of skin surface temperature comes from the concept of a central hot 'core' from which heat emanates to the surrounding tissues. This leads to a distribution where the temperature decreases as the distance from this 'core' increases. However, as we shall see this only gives a partially accurate description for a person at rest in moderate and warm environments. In hot conditions and during exercise where there are local areas of heat production, or in situations where there is considerable sweating, this pattern is greatly modified.

Measurement of skin temperature

The traditional methods for measuring skin temperature have been to use thermocouple or thermistor probes attached to the skin surface; in order to build up a comprehensive picture of the skin temperature distribution over the whole body, measurements at many different sites are necessary. This involves simultaneous readings from a number of probes with a complicated array of wires leading from the sensors to the measuring instruments. Consequently, these readings can often be unreliable.

This problem makes for difficulties of measurement in all conditions except where the subject is lying or sitting in the same position for the whole period of observation. Out of doors, on an active subject who is running, for instance, such methods are extremely difficult. For this reason other techniques have been developed, specifically telemetry. Temperature signals can be picked up from various sites on the subject and then transmitted to a receiver at some distance. This can be a very effective technique and has been widely used. The problems are that it can be difficult to follow more than one subject at a time, due to interference, and the range of such telemetric signals is usually limited to about 0.5 km.

Recent advances in portable cassette tape recorders have enabled signals from temperature sensors on a subject to be recorded for as long as 24 hours. Such recorders may have up to six channels on which to store individual temperature readings and when these recordings are played back after the period of observation, a continuous record of the temperatures at all the sites is produced which may be fed into a computer for statistical analysis.

One of the difficulties of measuring skin temperature is that out of doors, or indeed indoors in some situations (in factories with furnaces for instance) if there is a high level of thermal radiation, surface temperature may not be measured accurately by a thermocouple or thermistor applied to the body. Such a probe has to be kept in continuous contact with the skin and this is usually done simply with a strip of sticking plaster. In some cases the thermistor or thermocouple may be shielded from radiation and may be cooler than the surface of the exposed skin. On the other hand, if the probe is not shielded, it is exposed to radiation and may reach a higher temperature than the underlying skin. In either case, when the skin is subjected to solar or other strong radiation the surface temperature can only be measured with some difficulty.

One method which has been used to try to overcome this in situations where there may be a very steep gradient of temperature on the outer surface of the skin, is to use a thermocouple mounted in a needle. The needle can be inserted into the part of the body under investigation and the point of the thermocouple brought upwards towards the skin surface enabling temperatures at various levels in the tissues to be measured. Once the skin has been pierced from below the thermocouple will record a lower temperature; this method is useful from the point of view of measuring the gradient through the tissues of the body in order to establish thermal conductivity coefficients.

Use of infra-red thermography
When the body surface is covered with clothing it is always necessary to use thermocouple probes, thermistors or sometimes resistance thermometers to measure skin temperature. However, for exposed skin or the outer surfaces of clothing the technique of infra-red thermography is available. In recent years, this has developed from being a military tool for the location of men and machines in total darkness, fog, etc. to being a powerful medical and physiological technique with numerous applications. As well as being used to study pathological conditions it is now proving extremely useful in studies of thermal regulation. The technique, which has been outlined in Chapter 2, has the merit that it is non-invasive and probes and electrodes are not attached to the body surface. This means that the subject is able to perform tasks and to move around unhindered.

The 'normal' thermogram
There is considerable individual variability in skin temperature distribution and the concept of a 'normal' pattern in quantitative terms is difficult to define. However, there are a number of common features that can be distinguished over most subjects and these can be illustrated by the thermogram of

a man, woman and child in an environment of 22°C seen in Plate 1. The boy in this plate was 7 years old and the adults were in their early thirties. The temperature colour code is shown along the bottom of the picture with black being 26°C and cooler, white 34°C and warmer and adjacent colours differing by 1°C.

Perhaps the most obvious feature of this thermogram is the range of temperature over the subjects. On the adult man this was more than 10°C, with the skin overlying the knees being the coolest at below 26°C; over the other two subjects the range was approximately 7°C.

A notable feature in many thermograms of lean subjects is the warm area over the top of the shoulders and spine and in Plate 1 it is more marked in the female. In adults, this area generally forms a 'T' or 'Y' shaped area but in children there is a larger proportion of the back at higher temperature in a 'V' shape extending from the shoulders to the coccyx.

There is a development of the thermal pattern from infancy and through childhood (see infant thermogram in Plate 15, and Chapter 11); once the adult pattern is established it appears to persist in healthy lean subjects and has been observed up to the eighth decade of life.

The classic concept of a temperature distribution based on the idea of a hot 'core' can be seen from these thermograms only to partly describe the complex pattern. The skin overlying the so-called 'core' (upper trunk and head) is generally warmer than that over the hands, arms and lower limbs. However, superimposed on a pattern described in this way are the effects of heat conduction through the tissues from thermogenic structures underlying the skin surface together with the influence of skin blood flow and the proximity to the surface of major blood vessels.

Plate 2(a) shows an example (in moderate to cool environments) of the decrease in temperature towards an extremity with, in this case, a symmetrical fall of more than 7°C over the hand from the palm to the fingers. A modification to a similarly decreasing pattern is seen in Plate 2(b) which shows the right foot from the medial side where the structures underlying the skin directly influence the surface temperature. The skin over the medial malleolus, where the thermal insulation is high, contrasts with the area over the posterior tibial artery which is more than 4°C warmer.

Clothing and hair are good thermal insulators and the thermograms in Plate 2(c), (d), (e) and (f) are of a subject (in fact author OGE) who is bald with only a little hair at the back of the head, in an environment at 22°C. The temperature distribution over the lower part of the face is continued over the upper half whilst the hair has a surface temperature some 5–6°C lower than the bare skin. It is interesting that temperature of the hair is very close to that of the outer surface of the clothing, in this case a jacket over a shirt. In this picture the pinna of the ear is a cool area but deeper, past the concha and into the external acoustic meatus, there is a higher temperature which is consistent with that measured by tympanic probes as an index of deep body temperature as discussed in Chapter 2.

Elsewhere, mention is made of the fact that it is difficult to see a good physiological reason for the differences between black and white skin. In this connection, it is of interest that the resting temperature patterns over black and white subjects of similar age and build show no difference other than the individual variations observed between subjects of one colour.

As already mentioned, there is considerable individual variation between subjects in actual skin temperatures and their distribution. Plate 3(a) shows an example of these differences in temperature over the faces of two subjects. Variability also occurs in any individual and this depends on factors such as environmental temperature, degree of activity, emotional state, etc. If standardized thermographic examinations are carried out in similar environments and with suitable equilibration, there is substantial repeatability of temperature and pattern on separate occasions with the same subject.

Plate 3(b) is a thermogram of a young woman who was seven months pregnant and is in marked contrast to the young woman in Plate 1. In pregnancy the whole of the upper trunk increases in temperature. This is in contrast with the abdomen which shows a cooling, possibly due to the thermal insulation of the amniotic fluid.

Rubefacients and skin temperature

The application of irritants to the skin is very old and was originally based upon the idea that internal disease consisted of a malignant 'tumour' which could be drawn to the surface by irritation. Fairly early, it was recognized that skin irritation often relieved pain and this principle is used in modern drugs which produce skin vaso-dilation, or hyperaemia, with very little actual irritation. These drugs are known as rubefacients and proprietary compounds are available which are used in the relief of rheumatic and minor injury pain. The effect on skin temperature of one such compound (an embrocation stick containing ethyl, methyl and glycol salicylate together with capsicin) is shown in Plate 3(c) and (d).

This rubefacient was applied over the subject's back in the form of a circle around a cross. The skin temperature over this mark increased steadily over a period of about 30 minutes and was accompanied by a feeling of warmth, and some irritation. The thermogram shows the circle and cross as white, indicating an increase of several degrees over the surrounding area. The temperature subsequently decreased over approximately the next 30 minutes as the skin surface cooled and returned to normal.

Cold stress tests

The vaso-constrictive response of peripheral blood vessels to a cold stimulus, and the hyperaemia associated with subsequent vaso-dilation, can be assessed by the effect that the changing bloodflow has on skin temperature. A standard test is for a hand, covered by a thin plastic glove, to be immersed in cold water (around 5°C) for 1 minute. When the hand is removed from the water and the glove taken off the rise in temperature of the skin can be monitored by thermography. Plate 4(a) shows a sequence of thermograms during such a test. The base and tips of the fingers re-warm first, with a hot 'track' joining the two areas. The central parts of the fingers subsequently re-warm from both sides.

With computerized analysis (Plate 4(b)) it is possible to plot a re-warming curve for this sequence of events in one finger and this is seen in Fig. 3.1(a) with Fig. 3.1(b) being a typical re-warming curve for a complete finger and whole hand.

The effect of environmental temperature

Skin surface temperature distribution is markedly dependent on the temperature of the environment. This is illustrated in Plate 5 which shows whole body thermograms in environments of 10°C (Plate 5(a)) and 38°C (Plate 5(b)). At the lower temperature the influence of thermogenic structures underlying the skin was more marked, with the temperature range over the subject in excess of 10–12°C.

At the higher ambient temperature the distribution was quite different. Passive areas, such as the hair, which do not have the ability to thermoregulate, closely followed the temperature rise that occurred in the ambient air and had the highest temperature. There were, in general, larger areas of skin at uniform temperature; this was due to a reduced gradient between the deep tissues and the environment, coupled with greater cooling because of increased sweating and evaporation. The temperature range over the body, which was 10–12°C in the cool can be reduced to 3–4°C in hot conditions.

In the cool, one of the warmest areas was the axilla (Plate 5(c)) due to the presence of major blood vessels below the skin surface. However, this area is also well endowed with sweat glands and in hot conditions (Plate 5(d)) had the lowest skin temperature because the evaporation of sweat greatly increased local cooling. The temperature range of 10–12°C over the face and chest in the cool also reduced to 3–4°C.

Lean and obese subjects

In any one subject, the skin temperature distribution can be influenced by local subcutaneous fat thickness but, in general, fat people have different overall temperature distributions compared with lean subjects. Plate 6(a) shows an example of skin temperature over an obese female. In the posterior view the warm 'Y' shaped area over the spine and top of the shoulders is partly 'obscured' by cooler skin and changed to a 'butterfly' shape. In the front view there is often a cape-like cool area over the shoulders and pectoral region which, like the cold areas over the back, may be due in part to the insulating effect of fat although it is notable that cold areas over fat upper arms are similar to those sometimes found in very lean subjects.

Plate 6(b) and (c) show the temperature pattern over an extremely lean subject suffering from a rare collagen disease known as Ehlers–Danlos Syndrome Type IV. In this condition there can be extremely thin skin (perhaps one-third of normal thickness) and practically no subcutaneous fat. Such patients could be said to provide an extreme contrast with obese subjects. The temperature distributions in such cases have many features associated with healthy lean subjects, and although temperatures over the face and trunk can be fairly high, the overall pattern does not appear to be dominated by direct conduction of heat from the 'core' as much as might have been expected.

Colour Plates Section
Captions

Plate 1
Thermograms of a man, woman and seven year old boy in an ambient air temperature of 22°C. Black 26°C and below, white 34°C and higher in steps of 1°C.

It is notable that the upper part of the body is hotter than the lower in all of the subjects with the skin over the knees of the adult man being a marked cool area. On this subject the range of temperature overall is greater than 10°C. The pattern over the back of the boy has the characteristic large 'V' shaped warm area seen in childhood. In the adult this warmed area persists but its shape is more of a 'T' or 'Y' and is more marked on the female subject in this case.

Plate 2
(a)   An example of temperature gradient in a hand with a symmetrical drop of some 7°C occurring over all of the fingers. This thermogram was taken under resting conditions at an air temperature of 22°C. Temperature patterns over the hands can be very variable with the blood supply to the fingers able to cause rapid warming. If patterns such as this exist under conditions when rewarming might be expected and when the temperature gradient is of the order of 9°C, then this can be indicative of a vaso-spastic condition such as Raynaud's phenomenon. (Black = 27°C, white = 35°C in steps of 1°C.)

(b)   Thermogram of a right foot from the medial side at an air temperature of 22°C. The skin covering the medial malleolus is cool; the thermal conductance through bone is low. This contrasts with the skin over the posterior tibial artery which is more than 4°C warmer. These features are superimposed upon a generally decreasing temperature towards the extremity of the foot. (Black = 27°C, white = 35°C in steps of 1°C.)

(c) and (d)   Thermogram of author, OGE who is bald (in an environment of 22°C). The temperature distribution over the lower part of the face continues to the upper part of the head. In the posterior aspect the bare skin is some 5–6°C warmer than the hair which has a temperature near to that of the clothing. (Black = 27°C, white = 35°C in steps of 1°C.)

(e) and (f)   Thermograms taken over a subject wearing a shirt, jacket and tie at an ambient temperature of 10°C. In the posterior view the intimate contact of the clothing with the shoulders causes a high temperature on the surface due to heat conduction through the fabric. On the front, the lapels and parts of the jacket not in direct contact with the body surface are cold with a temperature little different to that of the surroundings. (Black = 25°C, white = 33°C in steps of 1°C.)

Plate 3

(a) An example of the individual variability between skin temperature patterns. The upper two thermograms are of an obese male and the lower two of an obese female. They were taken at an ambient temperature of 22°C after a 10 minute equilibration period. Over the male the coolest parts of the face are the cheeks which are black; white warm areas around the orbit and temple indicate a temperature range of more than 8°C.

For the female, the cheeks are the warmest part of the face and differ by some 10°C from those of the male subject. (Black=27°C, white=35°C in steps of 1°C.)

(b) Significant temperature changes occur during pregnancy. This thermogram of a young woman seven months pregnant is in contrast to the temperature distribution seen in Plate 1. The whole of the upper part of the trunk, including the breasts and shoulders, shows a marked overall temperature increase. This is presumably associated with the development of the breasts. There is a marked cut-off leading to a cooler region over the abdomen which may be due in part to the insulating effect of the amniotic fluid surrounding the foetus. (Black=29.5°C, white=34°C in steps of 0.5°C.)

(c) and (d) An example of the way in which a chemical irritant can modify skin blood flow and therefore surface temperature. A rubefacient (a preparation containing ethyl, methyl and glycol salicylate and capsicin) was applied to the back of the subject in the form of a circle and cross. After some 30 minutes the thermogram showed a marked increase of temperature associated with the very obvious reddening of the affected areas shown by the ordinary photograph (d). (Black=32.5°C, white=36.5°C in steps of 0.5°C.)

Plate 4

(a) Cold stress tests are useful in assessing the competency of the peripheral vascular system. This series of thermograms (taken at 20 second intervals) illustrates the thermal recovery of a hand (covered with a thin plastic glove) immersed in water at 10°C for 1 minute. The base and tips of the fingers re-warm first with the central parts of the fingers warmed from both sides. Fig. 3.1 shows these results numerically. (Black=29°C, white=33.5°C in steps of 0.5°C.)

(b) An example of a thermal image that has been digitized and processed by an image analysis computer. The computer has produced a range of 15 colours to cover the temperature over the subject and has been programmed to assign specific temperatures to each colour.

Plate 5

An example of the influence of the environment on skin surface temperature.

(a) Thermograms of a subject taken at an air temperature of 10°C. There is a wide thermal range over the skin surface, with more than 10°C difference on the face between the cold nose and warm areas of the orbit and forehead. (Black = 26°C, white = 34°C in steps of 1°C.)

(b) The same subject exposed to an air temperature of 38°C. The temperature range over the skin is reduced, particularly over the face, with large areas at more uniform temperature. Surfaces such as the hair, which cannot thermoregulate, take up the temperature of the surroundings and appear as white and hot in this high ambient temperature. The patterning on the background is due to the walls of the experimental chamber heating up as the air temperature increased. (Black = 29.5°C, white = 37.5°C in steps of 1°C.)

(c) and (d)   Temperature patterns over the axilla at the two environmental temperatures show that in the cold (c) this is one of the warmest areas due to the proximity of major blood vessels. However, in the hot conditions (d) evaporation leads to it being the coolest area on the front of the body.

Plate 6

(a) An example of the skin temperature patterns found over obese subjects. These thermograms are of an obese female aged 19 years taken after 20 minutes equilibration at an environmental temperature of 26°C. A marked feature is the 'cape' like cold area extending over the shoulders and breasts and continuing to the posterior view where there is a significant modification to the expected warm area over the spine. (Black = 30°C, white = 34.5°C in steps of 0.5°C.)

(b) In the rare condition of Ehlers–Danlos Syndrome Type IV, patients have an extremely thin skin (perhaps one-third of normal) and practically no sub-cutaneous fat. Such patients provide an extreme comparison with obese subjects (Black = 31.5°C, white = 36°C in steps of 0.5°C.) as in these thermograms.

Plate 7

(a) Thermograms showing temperature changes that occur during exercise. The subject, an Olympic class marathon runner, was running on a treadmill in a climatic chamber at an air temperature of 10°C. (Black = 20°C, white = 28°C in steps of 1°C.)

    (i) Temperature pattern over the athlete at the start of running at 16 km/h (the white band on which he appears to be running is the area on the treadmill heated up by his feet).

    (ii) Temperature distribution over the athlete after some 15 minutes running showing an overall cooling of the body surface.

    (iii) After approximately 25 minutes running, temperatures had fallen further due to the onset of sweating and evaporative cooling. Just before this sweating there had been a marked increase in the temperature of the hands and forearms but this disappeared due to the evaporative cooling. These experiments were carried out both in 'still' conditions and with the runner subjected to a wind at his running speed.

(b) A series of thermograms showing the 'heat surge' as the athlete stopped running. Temperatures immediately began to rise over the skin surface to produce extensive areas of white indicating increases of at least 10°C in some areas. The eight thermograms in this figure represent a time of approximately three minutes. (Black = 20°C, white = 28°C in steps of 1°C.) When thermograms were complemented by temperature and heat transfer measurements (using thermocouple probes and heat transfer calorimeters attached to the skin) the heat surge was seen to extend over the thighs for some six minutes as illustrated in Fig. 3.2.

Plate 8

(a) Thermographic investigations were made out of doors on a running track at an environmental temperature of 20°C with the Olympic marathon runner together with a club standard athlete. Similar changes to those observed in the chamber experiments were found over both athletes and (a) shows the distribution in the steady state. The Olympic runner on the left of the picture shows a marked hot white area over the heart muscle in contrast to the black area over the abdomen and chest which is at least 10°C colder. (Black = 23°C, white = 31°C in steps of 1°C.)

(b) In this investigation the club athlete had to retire prematurely with anhydrotic heat exhaustion and (i and ii) show the increase of temperature that occurred due to his inability to sweat and lose heat by evaporation to the environment. There was an overall temperature rise of 10–12°C in 1–2 minutes. (iii) shows the well-controlled thermoregulation of the Olympic runner at the same time as the heat exhaustion occurred in his companion. (Black = 23°C, white = 31°C in steps of 1°C.)

(c) An example of temperature changes observed over a subject riding on a bicycle ergometer having moveable handle bars for arm exercise. A well-defined hot area of skin over the active muscles of the legs occurred after some 10 minutes exercise. Hot areas over the joints and muscles of the arm and elbow are also evident. (Black = 26°C, white = 34°C in steps of 1°C.)

Plate 9

A series of thermograms taken of a subject in a room temperature of 25°C over a period of four hours. The subject lay uncovered on a bed. During this period there were intermittent periods of sleep and the investigation was terminated after four hours when cold discomfort precluded further sleep. A notable feature is the high temperature over the face and head. During this time, the hands underwent rhythmic temperature changes and were markedly warm at the end of four hours. (Black = 31.5°C, white = 35.5°C in steps of 0.5°C.)

Plate 10

(a) Thermograms of the head of the subject shown in Plate 9 taken at a different sensitivity to show thermal patterns over the face. Mean temperatures for the head were evaluated and plotted in Fig. 3.3 to show the rhythmic nature of the changes. (Black = 28°C, white = 36°C in steps of 1°C.)

(b) Thermograms of the left foot of a second subject undergoing thermography during sleep. (Black = 31.5°C, white = 35.5°C in steps of 0.5°C.)
   (i)   shows the temperature distribution with the subject asleep. (Mean temperature 34.1°C).
   (ii) The temperature pattern immediately after (i) with the subject awake showing a mean temperature of 33.7°C.
   (iii) Temperature patterns some 30 minutes later with the subject again asleep showing exactly the same distribution as in (i) (Mean temperature 34.1°C.)

There appears to be a quick acting response in skin temperature to periods of sleep and wakefulness.

Plate 11

An example of the rhythmic temperature changes observed over a young female subject in a calorimeter chamber with a well controlled environmental temperature of 26°C. The observations were initially made in an effort to visualize any thermally active so-called 'brown fat' areas underlying the skin of the back. In this case, the thermograms were recorded at 15 minute intervals over a two and a quarter hour period before and after administration of the drug ephedrine (60 mg taken orally) which had been thought to be able to initiate brown fat thermogenesis. Instead of observing any very localized thermogenesis, large areas of the body were found to be involved in subtle changes as can be seen from this sequence. Large areas on the front of the trunk showed a general cooling for some 60 minutes. This was followed by warming for the next 40 minutes and then slight cooling for the last 35 minutes.

In contrast, the back started cooler than the front and warmed generally, particularly over the neck and spine, for some 75 minutes. Cooling then occurred for some 30 minutes, followed by slight rewarming. A similar rhythmic pattern was seen with this subject (although not quite so marked) after a drink of 300 Calories. (Black = 32°C, white = 36°C in steps of 0.5°C.)

Plate 12

Examples of the use of infra-red thermography in clinical assessment and diagnosis.

(a)   A thermogram of the left hand of a 14 year old boy complaining of coldness and numbness in the index finger. The thermogram shows a clearly defined cold area apparently limited to this finger and it was initially thought that this might have resulted from some infection or injury to the nerve or blood supply.

     When a cold stress test (immersion of the hand, covered in a plastic glove, in water at 10°C for 60 seconds) was performed the temperature recovery time (plotted in Fig.3.7) was seen to be sluggish compared with a healthy child. At the time that the affected hand was immersed in water there was a contra-lateral response and the index finger of the right hand showed a marked cool area. The conclusion from these tests was that the overall thermal response of the two hands was abnormal and that this was not a local injury or infection but indicative of a more generalized involvement of the vaso-motor system. (Black = 31°C, white = 35.5°C in steps of 0.5°C.)

(b)   A thermogram of inflammation over an elbow joint associated with rheumatoid arthritis. The temperature pattern over such an inflamed joint is characteristic and different to the pattern where larger areas of soft tissue may be involved in the inflammatory process. (Black = 29°C, white = 33°C in steps of 0.5°C.)

(c)   A thermogram showing extensive hot areas of inflammation over both knees of a four year old child suspected of suffering from juvenile chronic arthritis. At the time of this examination the left knee was clinically affected with pain and swelling but the right knee showed no symptoms. Several days following this examination the patient was readmitted to hospital with pain and swelling in the right knee. The thermographic examination had clearly been able to pick up an inflammatory condition when sub-clinical. (Black = 31°C, white = 35.5°C in steps of 0.5°C.)

(d) and (e)   A thermogram and visible light picture of an ulcerated area on an elderly patient suffering from vascular insufficiency in the legs. The ulcerated area itself is cold with the skin of the foot distal to the affected area at least 5°C warmer. Such cold ulcers seem to be associated with vascular disorders and healing tends to occur in a direction from the hottest unaffected area towards the coldest regions. (Black = 29°C, white = 33.5°C in steps of 0.5°C.)

(f)   Chemical or surgical sympathectomy is sometimes performed on patients with vascular insufficiency who have developed ulcers on the legs. This is an attempt to increase the blood supply to the affected area and so facilitate healing. This thermogram shows the effect of sympathectomy, performed some two months previously, on the right leg of an elderly patient. The right leg is some 1–1.5°C warmer overall than the left. Thermography is a useful clinical tool in assessing the effect of such sympathectomy. (Black = 29°C, white = 33.5°C in steps of 0.5°C.)

(g)   A thermogram of the legs of a middle-aged diabetic female with a long-standing ulcer on the right foot. This patient did not have arterial disease of the legs and the blood flows according to Doppler ultrasound techniques were symmetrical and within normal limits. The pulses at both ankles were also normal. The notable feature of this thermogram is the thermal amputation showing the left lower leg and foot below the temperature of the examination room (26°C) during the hour-long investigation. Current research is aimed at elucidating the mechanisms involved in such conditions. The ulcers associated with diabetic neuropathy also seem to have a tendency to be in hot areas in contrast to those ulcers associated with vascular insufficiency. (Black = 26°C, white = 30.5°C in steps of 0.5°C.)

Plate 13

(a) In extensive eczema, because of the very large surface areas involved, there can be an impairment of thermoregulatory function. Even at room temperatures of up to 28°C there may be shivering. Thermography is useful at identifying the eczematous areas and monitoring their recovery. The affected skin of such a patient is shown in thermograms i and ii. There was a notably increased temperature over the arms and anterior trunk, but in addition there was a changed distribution over the back of this subject during the acute phase. The pattern is seen to be unstructured but less affected by temperature rise than the arms and rest of the trunk (i). The mean temperature of the back in this condition was 34.2°C. After treatment with steroid drugs and with the patient living in an air-conditioned allergy free room substantial recovery took place in about one week. The thermal pattern during recovery was seen to be much nearer to the normal for the back (ii), with residual eczematous patches still at higher temperatures. The mean temperature during recovery was 33.4°C although patches up to 36.4°C were observed. (Black = 31°C, white = 35.5°C in steps of 0.5°C.)

(b) Sunburn, even in mild form, can modify skin temperature patterns as illustrated here. (i) is a thermogram of the back of a patient (incidentally having reflex sympathetic dystrophy) the day after a mild sunburn with the skin obviously red but not painful. (ii) shows the pattern one week later with the expected and charateristic 'V' pattern clearly evident. (Black = 31°C, white = 35.5°C in steps of 0.5°C.)

Plate 14

(a) Thermograms showing the temperature pattern over a subject wearing cold weather clothing during studies of garments suitable for use beneath hovering helicopters on ships at sea in the arctic as described in Chapter 10. Thermograms were recorded in still air conditions and also in the presence of helicopter downdraught. This illustration shows hot, red areas where heat was 'leaking' through the fabric. A number of garment assemblies were compared in this way and it was possible to choose optimum clothing where this heat leakage was least. The thermogram also shows the relatively high temperatures over the unprotected face, which in these conditions, and with this level of clothing, can account for up to some 40 per cent of total body heat loss. (Purple = 8°C, white = 20°C in steps of 2°C.)

(b) A thermogram of a seven year old boy and his dog, illustrating several important physiological principles. The face of the dog is covered with hair and has an overall temperature similar to that of the boy's hair and clothing. The boy's uncovered face has a high temperature and the contrast with the temperature of the hair and clothing illustrates their effectiveness as insulation. The cold black nose on the dog is because it was wet and losing heat to the environment by evaporation. The right eye of the dog was hot and this was found to be due to an infection which was subsequently treated. (Black = 29°C, white = 33°C in steps of 0.5°C.)

Plate 15
(a)   Thermograms of an infant nursed in a temperature-controlled incubator.
   (i)   The thermal pattern with a thermoneutral incubator air temperature of 32°C. The pattern shows relatively small variations over the body surface and no features due to subcutaneous thermogenic structures.
   (ii) The temperature pattern over the same infant when the incubator air temperature was reduced to 27°C. Subtle changes occur with the skin cooling over the back, towards the extremities and on the cheeks. The area over the carotid sheath shows an increase of temperature due to direct heat conduction from the major blood vessels.

(b)   A series of thermograms (taken over an 80-minute period) of an infant nursed in an incubator at 32°C, showing rhythmic changes similar to those observed in adults and described in Chapter 3 and Plate 11. Inset in this picture is an example of a computerized thermal image used for determining mean temperatures of the body, trunk and head and the results of such an analysis for this sequence of pictures is shown in Fig. 11.7. (Black = 32°C, white = 37°C in steps of 0.5°C.)

Plate 16
(a)   Composite Schlieren photograph of the convective boundary layer flow over a standing, nude subject at a room temperature of 22°C. The boundary layer reaches a thickness of some 15–20 cm at face height and extends over 1.5 m above the head. The maximum velocity of the flow over the face is between 0.3 and 0.5 m/s. The inherent limit of sensitivity of the Schlieren system generally means that only some half of the total boundary layer thickness is seen.

(b)   An example of the interaction of boundary layer flows between subjects at close quarters! The potential for micro-organism transport and transfer between boundary layers is the basis of an important mechanism in airborne cross-infection.

(c)   Schlieren photograph showing the streaming convective airflow from the hands and fingers.

(d)   The expiratory flow from the nose and mouth can be clearly identified in the Schlieren system. The expired flow is generally taken away upwards by the airstreams moving past the face with little chance of inhaling any expired air by the next inspiration.

Plate 1

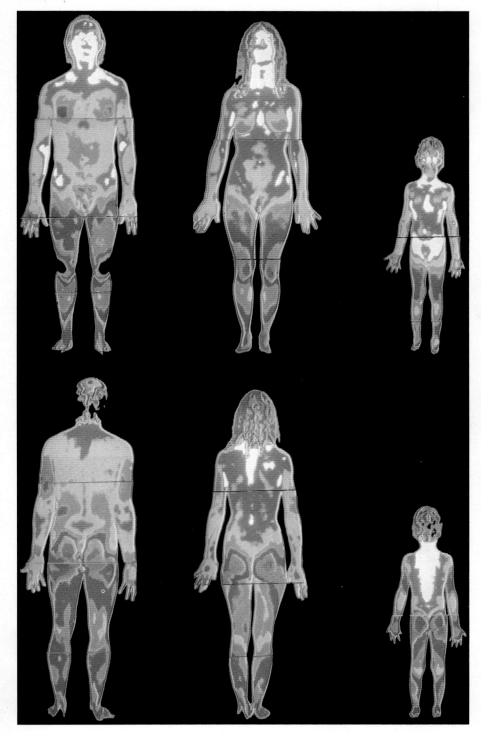

Plate 2

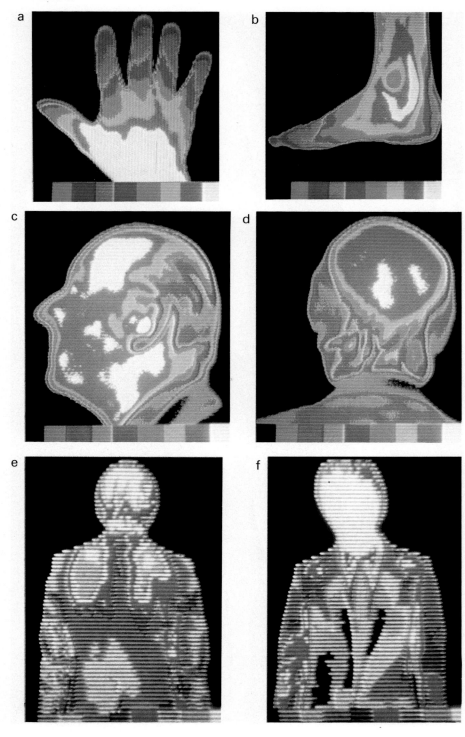

a

b

c

d

e

f

Plate 3

a

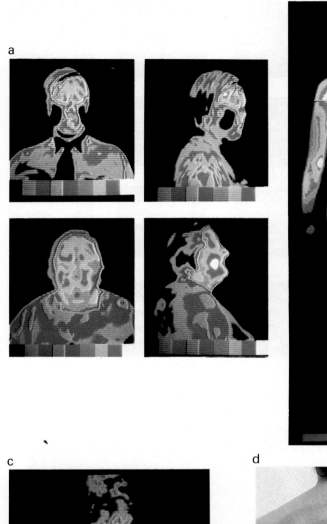

b

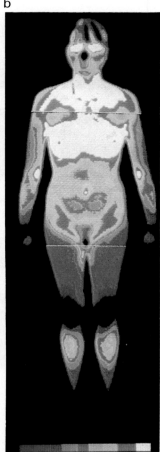

c

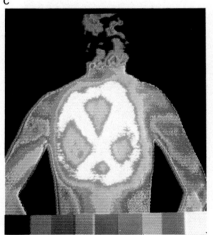

d

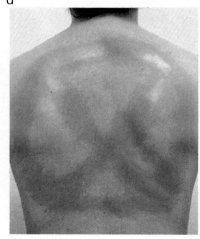

Plate 4

a

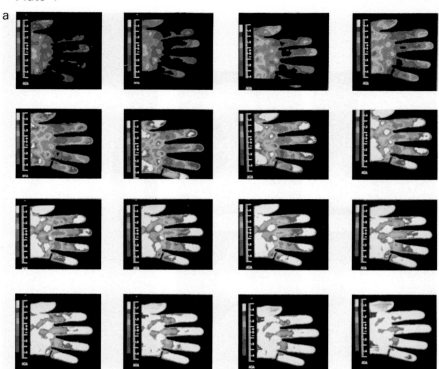

b

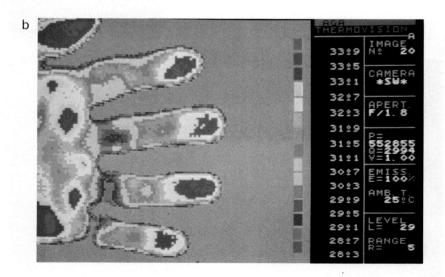

Plate 5

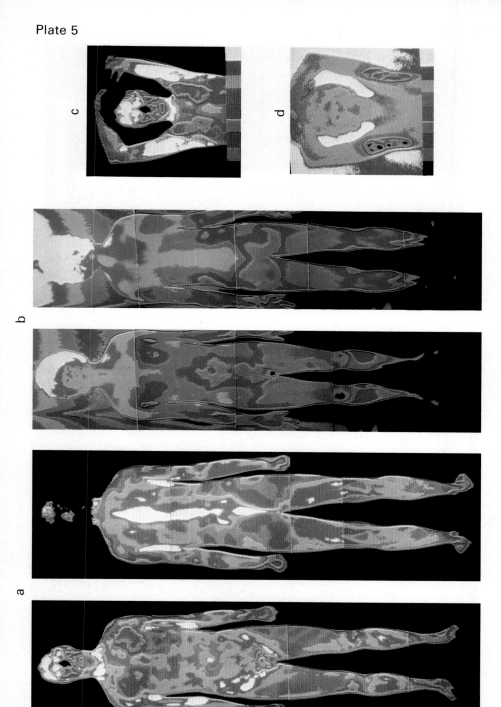

Plate 6

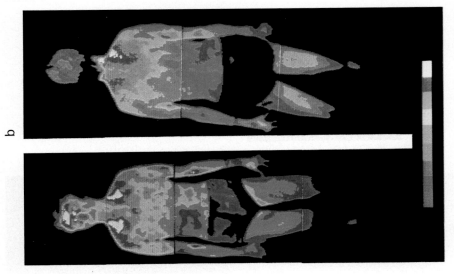

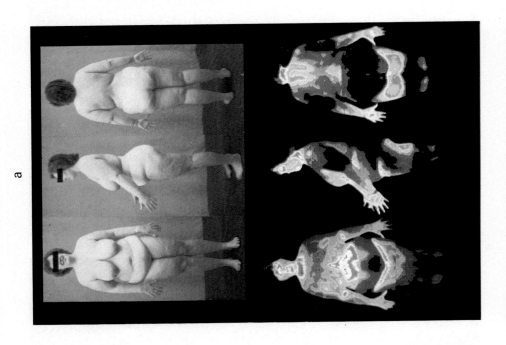

Plate 7

i ii iii

a

b

Plate 8

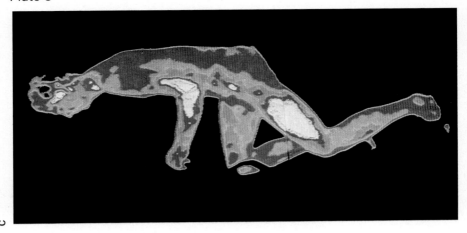

c

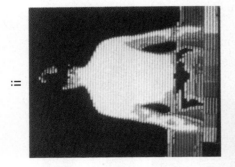

ii

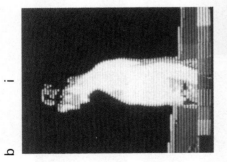

i

b

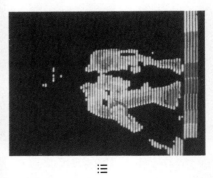

iii

a

Plate 9

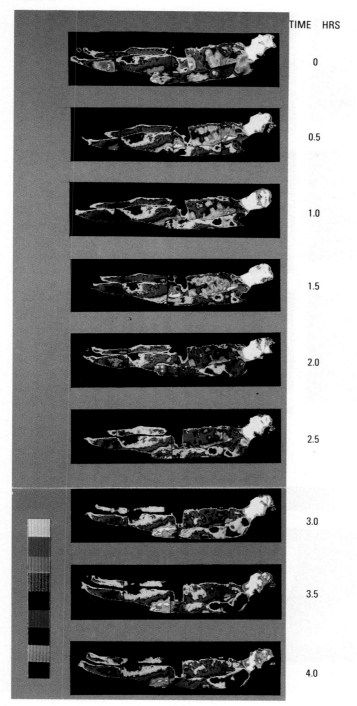

TIME   HRS

0

0.5

1.0

1.5

2.0

2.5

3.0

3.5

4.0

Plate 10

a

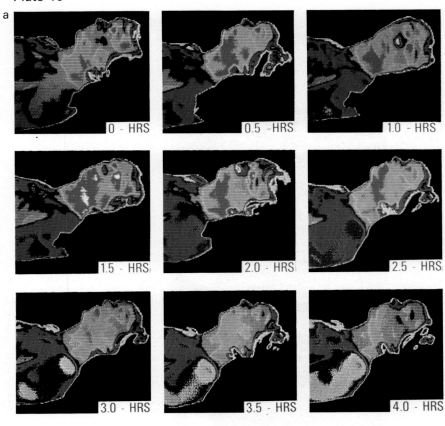

0 - HRS   0.5 - HRS   1.0 - HRS

1.5 - HRS   2.0 - HRS   2.5 - HRS

3.0 - HRS   3.5 - HRS   4.0 - HRS

b

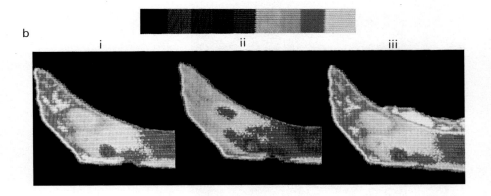

i   ii   iii

Plate 11

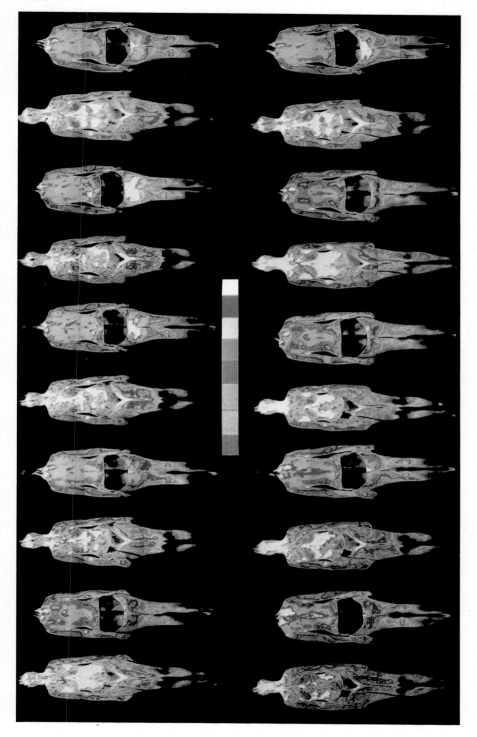

Plate 12

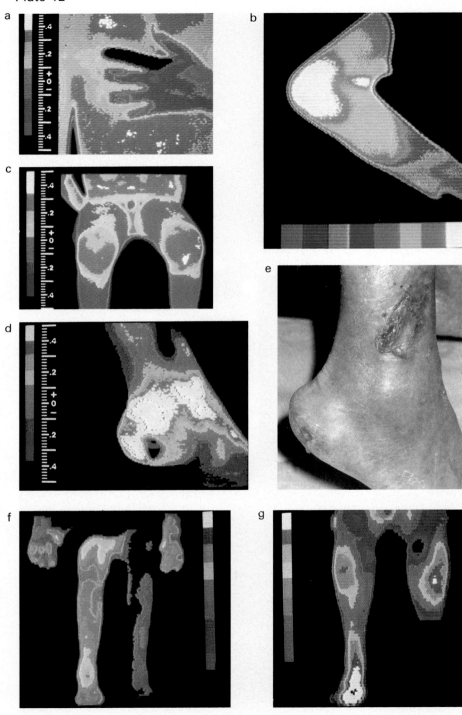

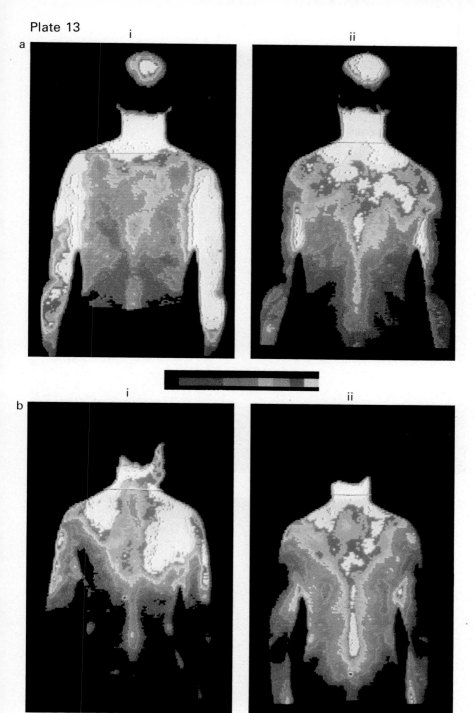

Plate 13

Plate 14

a

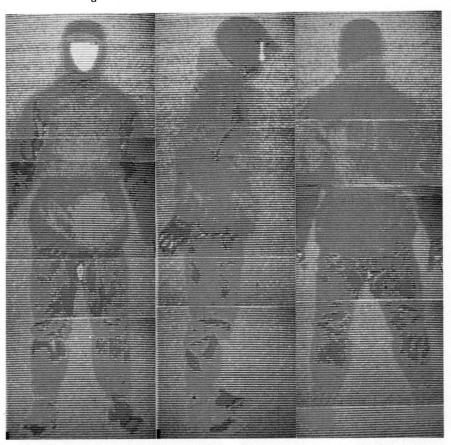

b

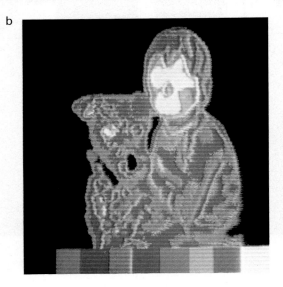

Plate 15

a

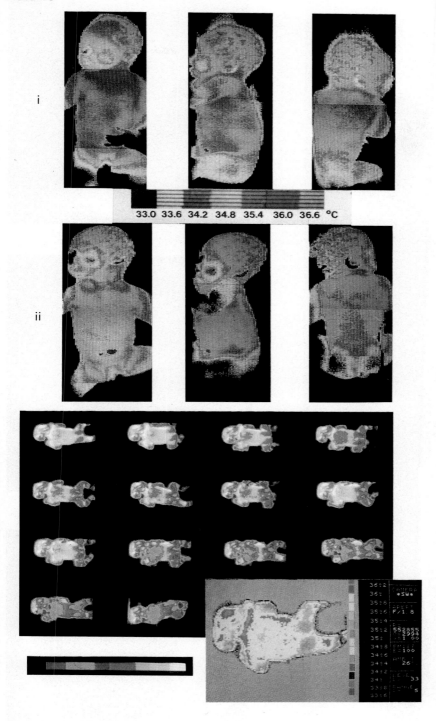

33.0 33.6 34.2 34.8 35.4 36.0 36.6 °C

Plate 16

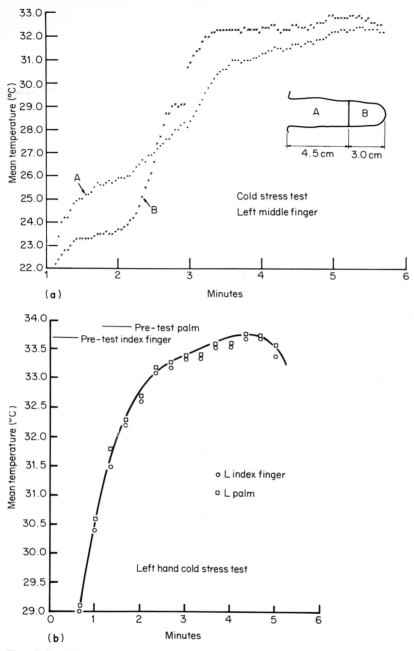

Fig. 3.1 Skin temperature re-warming curves for the finger (a) of one subject and the finger and palm of a second subject (b) after the hands have been subjected to cold stress by immersion (in a plastic glove) in water at 10°C for 1 minute. (a) shows an initial re-warming to the base of the finger followed by a rapid warming of the finger tip presumably due to the opening of arteriovenous anastomoses.[19] (b) shows that temperature recovery is complete within 4–5 minutes for a healthy subject. Extended recovery times are indicative of disease in the peripheral vasculature and such curves are a useful diagnostic aid.

Exercise and skin temperature

Activity raises body temperature although it is necessary to work extremely hard in order to raise it to 40°C. A marathon runner can achieve such a temperature and at the end of one race[1] temperatures in the rectum of the competitors were measured. Afterwards it was possible to list the competitors in the order in which they completed the race according to their body temperature. The winner had the highest at nearly 41°C; the athlete who came in second, a little lower and those who came in near to the end of the field had temperatures of 38°C or less. It is of interest to try and correlate these observations with temperatures which occur on the skin surface during exercise in cool or moderate environments.

Experiments to visualize the temperature distribution over the skin of a runner both indoors on a treadmill, and out of doors on a running track, have demonstrated very large changes[2]. Increased muscle metabolism, changes in skin bloodflow, and variations in environmental cooling all combine to produce a completely different temperature distribution during exercise compared with that at rest.

Plate 7(a)(i) shows the temperature over an Olympic class marathon athlete at the start of running at 16 km/h at a temperature of 10°C in still air. The subject, aged 31 years, was 179 cm tall, weighed 65 kg and had a mean skinfold thickness of only 4.18 mm.

Skin temperatures immediately began to fall when the subject started running, and reached steady values within about 15 minutes, having fallen some 5°C overall (Plate 7(a)(ii)).

Skin over the active muscles appeared 1–4°C warmer than at adjacent areas, especially over bone. When the subject had been running for approximately 25 minutes (Plate 7(a)(iii)) he began to sweat and this was marked by a further fall in temperature over the trunk and limbs. Colour reappeared over the hands and fingers (indicating a rise of 5–8°C) immediately before the onset of sweating and this temperature rise was complete within 2 minutes. Colour also reappeared over the forearms, but this re-warming was more gradual.

When wind was turned on to the runner at the same velocity as the treadmill, in order to mimic outdoor conditions, there was a further immediate fall in skin temperature of approximately 4°C but with little change in overall distribution. Temperatures continued to fall for about a further 20 minutes until the steady state was again reached, by which time many areas on the thermogram appeared black, especially over the abdomen. The skin over active muscles continued to be warmer than over other structures.

During such exercise, when the steady state had been reached, skin temperature distribution remained reasonably constant when heat production by the body and the rate of dissipation to the environment were equal. However, when exercise ceased, environmental cooling (due to movement of the body through the air) stopped immediately, whereas the overall heat generation of the body, and the muscles in particular, did not stop so abruptly but continued to influence skin temperature for some time afterwards.

This 'heat surge' was demonstrated in the Olympic athlete during the climatic chamber experiments. Plate 7(b) shows a series of thermograms taken at 30-second intervals after stopping running. The high initial temperature over the active muscles was seen to diffuse outwards as the whole skin temperature rose by several degrees. These thermograms were complemented by temperature and local heat transfer measurements and Figure 3.2(a) shows the time course of three temperature probes and Figure 3.2(b) one heat transfer calorimeter for 15 minutes after running had stopped. The temperature measurements reflected the thermographic visualizations and showed an initial rise for the anterior, medial and lateral thigh positions during the first 2 minutes after stopping running. It was only after approximately 10 minutes that the temperatures began to approach their final values. The heat transfer measurement on the anterior thigh showed an increase in heat loss for 5–6 minutes before falling steadily to reach an equilibrium near to the original value.

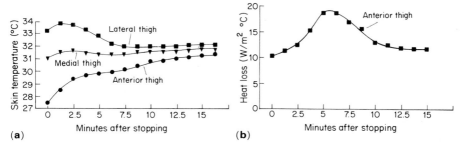

(a)   (b)

Fig. 3.2 Skin temperature (a) and heat loss (measured with a surface calorimeter) variations (b) over the thigh of an athlete for 15 minutes after stopping running. The heat 'surge' is consistent with the increased temperature seen thermographically and illustrated in Plate 7(b).

There is here an analogy between the human body and the motor car; when a car stops the cooling of the engine and radiator, due to the motion of the vehicle through the air (which together with the action of the fan provides forced convective cooling) ceases. However, the hot engine, having a large thermal mass, requires time to dissipate its heat. The result is that the water in the cooling system actually increases in temperature for some time after the car has stopped in much the same way as the skin temperature rises over the athlete.

A further series of experiments was carried out on a running track in a warmer environment of 20°C. Here the athlete 'competed' against a good club standard man of the same age and approximately the same height and weight but who had a mean skinfold thickness of 7.1 mm compared with the 4.18 mm of the Olympic man. Plate 8(a) shows a steady state temperature distribution during running with the Olympic man appearing on the left of the picture. There are many similar temperature features over the two athletes although the structures underlying the skin surface are more clearly defined over the Olympic man, probably because he was much thinner. The exercise again produced a complete redistribution of the thermal patterns seen at rest and, as in the chamber experiments, temperatures over some parts of the body fell by as much as 7–8°C. Skin over the active muscles was the warmest and only some 1–2°C lower than before running.

During this experiment, the club man had to retire prematurely. He suffered anhydrotic heat exhaustion, became unable to sweat and consequently overheated due to an inability to lose heat to the environment by evaporation.

Plate 8(b)(i) and (ii) show the effect that this had on skin temperature which rose by more than 10°C to give him an almost completely white appearance on the thermogram. During this time, the Olympic athlete continued to thermoregulate satisfactorily and was able to run without distress whilst maintaining the temperature distribution shown in Plate 8(b)(iii).

An example of temperature changes due to a lower exercise level is seen in Plate 8(c). Here, a subject was riding on a bicycle ergometer which also had moveable handlebars for arm exercise. The heat generated by the leg muscles became well-defined as a warmed area and appeared some 10 minutes after the start of the exercise. Hot areas over the joints and muscles of the arms and elbows can also be seen.

Skin temperature 'cycling'

On many occasions when skin surface temperature measurements are required it is often assumed that a 'steady state' exists when energy exchange with the environment and heat production stabilize to produce a skin temperature distribution which does not vary with time. For instance, as we have seen with large changes in environmental conditions or exercise rates, 20–30 minutes can elapse before skin temperatures approach anything like equilibrium values.

Recently, however, thermography has shown that there are more subtle temperature changes occurring over the whole body even when at rest and in well-controlled environmental conditions. During rest and sleep[3] the thermograms shown in Plate 9 were obtained (with a sensitivity of 0.5°C between adjacent colours). In this investigation the subject lay uncovered on a bed for a period of 4 hours in a room temperature of 25 ± 0.5°C. During this period whole body thermograms were taken at half-hourly intervals.

In this situation, the head and face of the subject were warmer than the rest of the trunk with a definite line of demarcation around the neck. The skin on the upper part of the trunk showed an alternate rise and fall in temperature for at least the first 2 hours. The hands and lower forearms, after an initial period of cooling, returned to temperatures approaching their original values whilst the shoulders and upper arms progressively cooled during the 4-hour period. At the end of this time a difference in temperature of 5–6°C existed between the shoulders and hands; the hands and face being the warmest areas on the body.

In order to investigate the temperature distribution over the head and face during this period a second series of thermograms was taken simultaneously with the first but at a sensitivity of 1°C between adjacent colours. These are shown in Plate 10(a) and reveal the temperature distribution over the face to be changing. When mean skin temperatures were evaluated from these thermograms and plotted against time, the result was as shown in Fig. 3.3. The

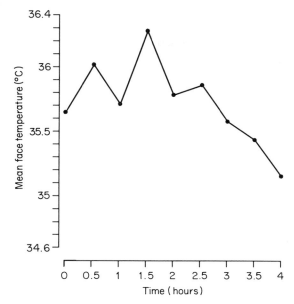

Fig. 3.3 Time variation of mean skin temperature over the face during a period of rest and sleep evaluated from the thermograms seen in Plate 10(a). There is a rhythmic change in temperature for some 2.5 hours after which a progressive cooling is seen.

face alternately increased and then decreased in temperature (by 0.2–0.5°C) for 2.5 hours after which time there was a progressive cooling. During this experiment the subject was dozing and not fully asleep except for brief periods; the observations were curtailed at the end of 4 hours as the conditions were then uncomfortably cool.

When this was repeated on a second subject substantially the same effects occurred although this subject was able to sleep part of the time. Plate 10(b) shows three thermograms of the foot and lower leg taken between 3 and 3.5 hours after the start of the experiment. In the first thermogram (i) the subject was asleep, and the mean temperature was 34.1°C. Immediately after this picture was taken the subject was awakened and a second thermogram (ii) of this area was made within approximately 30 seconds of the first. This showed an immediate fall in skin temperature of approximately 0.4°C presumably occasioned by vaso-constriction of the skin blood vessels on awakening. The subject then fell asleep again and half an hour later the thermogram produced an almost identical picture to the one taken when previously asleep.

Henane *et al.*[4] carried out a series of studies to determine the body and skin temperature and evaporation rates during sleep. In neutral conditions (32–34°C) body temperature and skin evaporation were found to decrease during the night and followed a rhythmic pattern. In warm conditions (up to 39.5°C) body temperatures and evaporation generally remained steady although in both environments fluctuations of body temperature and evaporation were seen to occur in association with rapid eye movement (REM) periods.

They found that mean skin temperature fluctuated continuously through the night synchronously with the occurrence of REM periods. Figure 3.4 shows an example of the continuous recordings of body temperature, skin temperature and weight loss during sleep and Figure 3.5 shows the phasic changes of local skin temperatures during REM sleep.

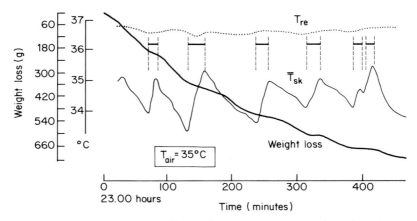

Fig. 3.4 Continuous recordings of body temperature (rectal and mean skin temperature T_{sk}) and weight loss during nocturnal sleep in a representative subject. Horizontal bars indicate REM sleep periods. (Reproduced with permission from Henane *et al.* (1977)[4].)

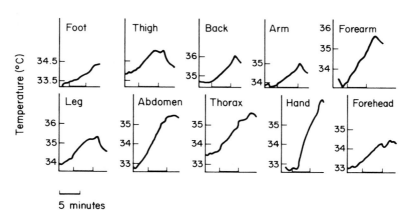

Fig. 3.5 Phasic increase of skin temperature observed during a REM sleep period in a representative subject ($T_{air}=35°C$). All plots are of the same REM period. (Reproduced with permission from Henane *et al.* (1977)[4].)

Skin temperature, obesity and brown fat

One of the factors that has been considered important in the mechanism of obesity has been to postulate areas of 'brown fat'. This tissue has been regarded as distributed at a number of sites over the body but mainly over the top of the back and over the shoulders[5]. Because brown fat is metabolically more active than surrounding tissue it was considered capable of raising tissue temperature and subsequently exchanging more energy with the environment. It was said that this mechanism was pronounced in some people and absent in others and that those with little or no brown fat could not lose as much energy to the environment as those with a greater supply. It was also thought that in those deficient in this thermogenic tissue the energy would manifest itself in the production of more fat cells and eventually in weight gain which would lead to obesity. This theory stated that it was not the total quantity of energy intake, in the form of food, that was responsible for weight gain but rather the ability or inability of particular individuals to lose their excess energy to the environment.

Whilst investigating this mechanism and attempting to confirm the existence of discrete 'hot spots' overlying these 'brown fat' areas in lean subjects a number of whole body thermograms were taken of which Plate 11 is one example. Temperature distribution at 15-minute intervals was observed after the administration of food and ephedrine (said to initiate 'brown fat' thermogenesis). The subject was a young female and the investigation was carried out in a calorimeter chamber with accurate temperature control in an environment of 26°C with an air movement of 0.2 m/s. The results revealed a temperature cycling similar to that observed during the rest and sleep investigation previously described. Significant temperature changes were seen over the whole of the front and back of the trunk. These appeared to be differently phased over the front and back after the administration of ephedrine but the pattern seemed to be rather more stable over the back when the subject had taken some food.

When these observations were repeated on a second subject, a male aged 50 years, similar changes were observed although after taking food the changes over the back of this subject were rather larger than over the front of the trunk. From these observations, the presence of small localized 'brown fat' areas over the back could not be substantiated. Large areas of skin were involved in continually changing temperature patterns over the body. From considerations of energy balance the heat exchange differences caused by these large areas changing temperature were in excess of those that could occur should only small 'brown fat' areas be thermogenic.

The exact quantification of the energy exchanges involved in these continually cycling temperatures has yet to be explored.

In a series of investigations of the temperature patterns over four obese females the mean temperatures for the front and back of the trunk and the face were evaluated. The subjects were each examined on four separate occasions under similar environmental conditions at 25°C and each subject completed a control session, a meal of 1200 J carbohydrate, an isocaloric protein meal and a 60 mg thermogenic stimulus of ephedrine[6].

The temperature changes over each 3-hour session are shown in Figure 3.6 together with average metabolic changes measured calorimetrically.

As in lean subjects the temperature changes that occurred following thermogenic stimuli were not confined to the 'brown fat' area. Except for perhaps Subject A with a protein meal, and Subject D with a glucose meal, the temperature changes observed in the obese subjects were rather less than in the lean people.

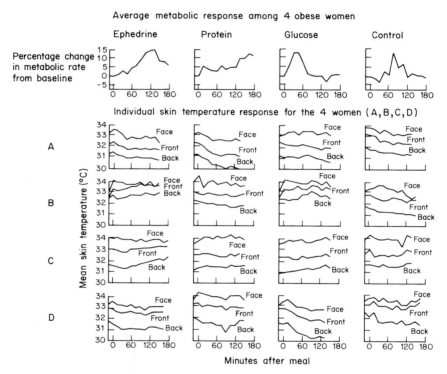

Fig. 3.6 Upper row of curves show average metabolic changes in four obese women over 3-hour periods as measured in a climatic chamber following various thermogenic stimuli. Rows A, B, C and D show mean skin temperature response for the four women measured thermographically over the face, front and back.

The influence of hot and cold drinks

The question often asked in hot summer weather is 'does a hot drink have more of a cooling effect than a cold one?'. Thermography was used to observe the temperature changes brought about either by an ice-cold drink or by a cup of hot tea[7]. Experiments performed in sunny conditions at a dry bulb air temperature of 31°C showed that when an ice-cold drink was taken there was little change in the thermographic picture except for some local cooling around the mouth. On the other hand, drinking a cup of hot tea produced a series of temperature changes which began some 2 minutes after drinking with an overall rise of 1–3°C in skin temperature accompanied by visible sweating. Five minutes after the drink, as the extra heat energy at the surface was dissipated by increased evaporation, radiation and convection, the skin temperatures fell by approximately 1°C from their pre-drinking values. Some 7 minutes after the drink, the skin temperatures had fallen a further 1°C and after about 9 minutes and another small fall, no further decrease was observed. The final skin temperatures over the upper trunk were 1–2°C lower than before the drink was taken. After a further 6 minutes temperatures slowly began to rise again and visible sweating stopped. The subject then said he felt cool as well as refreshed and dry.

This experiment was repeated on a number of occasions at different environmental temperatures with similar results; the effect lasting approximately 15 minutes.

Subsequently[8], further experiments were carried out using hot and cold water and coffee. The cold drinks produced no effects other than local cooling around the mouth and neck of up to 2°C; there was no general change in body skin temperature. With hot coffee, decaffeinated coffee and water, increased sweating was observed which again led to a fall in temperature. There were minor differences in the time-scale of these changes and it seems likely that the response was to the temperature of the hot fluid rather than to any biochemical stimulus.

The answer to the original question is that a hot drink on a hot day is indeed more effective at cooling the body than a cold drink.

Gustatory sweating, similar to that observed in these studies, has been reported before in response to chewing chillies[9, 10] and similar changes can be observed thermographically after eating a curry meal.

Skin temperature in clinical assessment and diagnosis

In a number of pathological conditions (locomotor, vascular and malignant diseases) skin temperature patterns frequently provide valuable diagnostic information and are useful in choosing, assessing and evaluating therapy. Amongst many uses for thermography[11] at present under investigation the identification of deep vein thrombosis is an area where screening has been successful and scrotal thermography is useful in assessing varicocele and male infertility.

Unfortunately, initial claims for thermography in the early detection of tumours associated with breast cancer were over-ambitious and brought the technique into some disrepute. However, with the advent of computerized thermographic equipment it is now possible to use skin temperature as a clinical index as long as it is related to a sound physiological background.

Plate 12 illustrates several clinical conditions where abnormal thermograms are found.

In vasospastic conditions, such as Raynaud's syndrome (a condition in which there is spasm in the arteries, arterioles and possibly the veins of the extremities), there is a characteristic thermographic pattern usually having, in the case of the hands, a distinct temperature cut-off at the wrists with a steep thermal gradient along the cold hands to the fingers. In such cases, when a hand is subjected to a cold stress test as previously described, the normal findings shown in Plate 4 and Figure 3.1 are greatly modified with, in some cases, only minimal re-warming occurring in periods of up to 1 hour.

In the early stages of such a condition thermal changes may not be so marked. Plate 12(a) shows the temperature pattern over the left hand of a 14 year old boy complaining of cold and numbness in the index finger. The steady state thermogram (at 25°C) showed a marked cool area over this finger. The fingers of the other hand had a normal temperature. However, when the left hand was subjected to a cold stress test the response in the right hand was a well-defined cool area which developed over the index finger.

The re-warming curve for the cold stressed left hand (and for the right hand tested separately) (Figure 3.7) showed a sluggish response compared with the control subject with the left index finger showing hardly any re-warming over a 20-minute period.

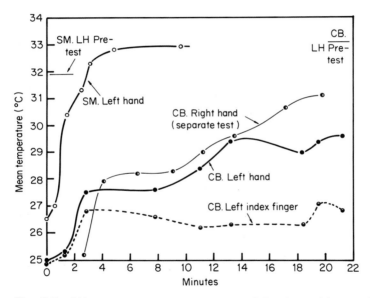

Fig. 3.7 Skin temperature recovery curves following cold stress for the patient (CB) having a cold index finger shown in Plate 12(a) compared with a healthy control (SM).

These tests, together with the steady state thermograms, indicated such a condition to be due to a generalized involvement of the nervous system, rather than to a local injury or infection as had at first been thought.

Thermography is particularly useful at identifying areas of 'inflammation', especially in rheumatoid arthritis where characteristic patterns are seen over inflamed joints (as in the elbow shown in Plate 12(b)). Numerical description of these patterns such as by mean temperature, area affected or thermographic

index[12] can be used to asses the effectiveness or otherwise of anti-inflamma-
tory drugs that may be used to treat such conditions.

Juvenile chronic arthritis (originally termed Still's disease) is considered as
a group of conditions with symptoms which include fever, rash, pain, swelling
and inflammation of joints together with soft tissue involvement. Thermographic patterns over such children are often abnormal in many respects and
their analysis can be useful in assessing the areas involved in individual cases.
Plate 12(c) shows the temperatures over the knees of a 4 year old boy in which
juvenile chronic arthritis was suspected. When this examination took place
there was clear clinical evidence of inflammation and swelling together with
pain in the left knee, but the right knee showed no symptoms. However, the
thermogram clearly indentified inflammation over both knees. Several days
later this boy was re-admitted to hospital with pain and swelling in the right
knee. In this instance thermography had identified inflammation at the sub-
clinical stage.

Blood perfusion in tissues surrounding ulcerated areas can be assessed by
thermography. Ulcers on the legs and feet are often associated with vascular
insufficiency and disease in major blood vessels. The degree to which an
ulcerated area is cold or 'dead' can be a useful guide to the potential for
healing. Plate 12(d) and (e) illustrate an ulcerated area near to the ankle of an
elderly patient with a poor blood supply to the legs. The affected area is cold
but is surrounded by tissue some 5°C warmer. In such cases, healing proceeds
gradually from the hottest unaffected areas.

Cold limbs are often found when the blood supply is reduced, for instance
in arteriosclerosis. This is illustrated in Plate 12(f) which shows a cold left
lower leg in an 86 year old arteriosclerotic woman. Initially, both legs were
similar but an ulcerated area developed near to the right ankle. This was
resistant to treatment and skin grafts were also unsuccessful. In an effort to
increase the overall blood flow to the leg a chemical sympathectomy was
performed some two months before the examination shown in Plate 12(f). It
appeared that this manoeuvre was successful at increasing blood flow and
raising temperature, with the right leg (mean temperature 32.2°C) being some
1.7°C higher than the left (mean 30.5°C).

Cold limbs are sometimes found even when there is no apparent reduction
of blood flow in major blood vessels. This is demonstrated in Plate 12(g)
which shows the temperatures over the lower legs of a 49 year old diabetic
woman. This patient had a neuropathy with insensitivity to touch and temperature stimuli on the front of both feet extending to the toes. She also had
a long-standing ulcer on her right foot. The insensitive feet of such neuro-
pathic patients can be particularly at risk of injury and infection which may
go unnoticed for some time.

At clinical examination this patient had apparently normal and symmetrical blood flow in both limbs as assessed by doppler ultrasound techniques;
she also had normal ankle pulses. The notable thermal feature of this patient
was that her left leg showed a sharp 'thermal amputation' just below the knee.
On many occasions this leg would remain below ambient temperature (25–
27°C) during periods of up to two hours at thermographic examination. In
addition, a computerized section taken longitudinally through the centre of

this cold leg showed a near linear fall of temperature with distance towards that part of the leg which was below ambient temperature (Figure 3.8).

Such features, whilst not common, are occasionally found and cannot be explained entirely satisfactorily by current theoretical heat transfer considerations.

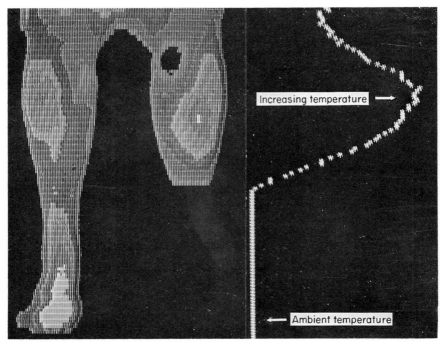

Fig. 3.8   Cold leg of the patient whose colour thermogram is shown in Plate 12(g). The right side of the figure shows a line scan through the cold limb with a near linear fall of temperature towards the lower leg. (Reproduced with permission from Clark, R.P. and Goff, M.R. (1984)[20].)

However, several points about available analyses can be made. The leg can be represented by a simple 'lagged' pipe model with blood flow only through central vessels and with the surrounding tissue being regarded as insulation or 'lagging' (Figure 3.9(a)).

In this model system there is a linear relationship (Figure 3.9(b)) between temperature and distance along the heat conductor and on the surface of the lagging; a pattern similar to that shown in Figure 3.8.

Heat transfer coefficients can be used to evaluate the quantities of heat exchanged with the environment from the lower legs of the patient shown in Plate 12(g). For this it is necessary to make approximations as to diameters, lengths and areas that will define cylindrical models representative of the legs. Figure 3.9(d) illustrates the results of such calculations with some 5 watts lost from the lower right leg and 2.6 watts from the upper part of the lower left leg. For the cold lower part of the left leg this calculation cannot be done as

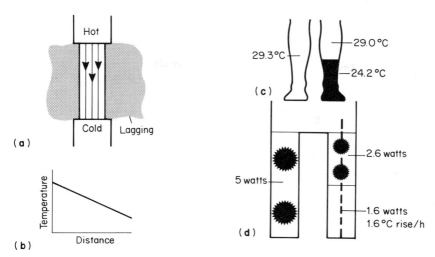

Fig. 3.9 A physical model representing central blood vessels and surrounding tissue as a 'lagged' pipe (a) and showing the linear relationship between surface temperature and distance (b). (c) shows the mean skin temperatures measured on the thermogram shown in Plate 12(g) and Fig. 3.8. (d) indicates the heat output from parts of the lower leg above ambient temperature. The lower part of the left leg, being below ambient temperature, does not lose heat to the environment from the skin but receives heat from the major blood vessels that should lead to a skin temperature rise of some 1.6°C/h.

the skin is below ambient temperature and does not lose heat to the environment. In this case if the model was considered to have a constant central temperature of 37°C (with no peripheral circulation) some 1.6 watts would be supplied to the leg and result in a temperature rise of the order of 1.6°C/h— a value greater than that found thermographically. With any blood available to the periphery this rate of rise would be more rapid.

The feet of this patient were subjected to hot (for the left) and cold (for the right) stress tests with the results shown in Figure 3.10. The left foot showed a cooling for some 6–7 minutes after hot stress before levelling out some 5.4°C higher than the pre-test temperature. The cold stressed right foot returned to the original temperature after some 12 minutes and continued to heat for the 20 minutes of the examination. This is within the range of recovery time seen in healthy controls and tends to support the clinical findings of normal blood flow to the limb.

Whilst currently available analyses are adequate to predict the relationship between heat transfer and surface temperature in most instances more research is needed into the mechanisms that may be involved in cases, such as just described, which cannot be adequately analysed.

Another area where thermographic evaluation is promising is in defining the extent of severe eczema. The eczematous areas are hot regions as can be seen in Plate 13. Extensive eczema is often associated with disturbances to thermoregulatory function frequently with coldness and shivering. In the

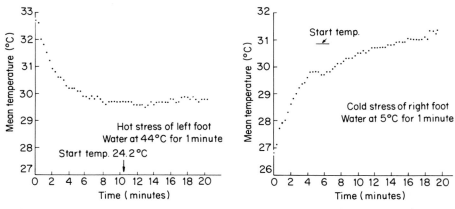

Fig. 3.10 Results of hot (left foot) and cold (right foot) stress tests for the subject shown in Plate 12(g) and Fig. 3.8.

example of Plate 13 skin temperatures greater than 36°C were found and there was also a change in the distribution of the thermal pattern. The usual 'Y' shape on the back disappeared during an acute period of eczema but returned during recovery.

Mean skin temperatures

As we have seen, skin temperature is not constant over the body and its distribution can vary with environmental conditions and with activity as well as by naturally occurring rhythmic changes. In calculations of heat transfer to the environment, and in the absence of computerized thermographic analysis or when temperatures beneath clothing are measured, it is necessary to evaluate an average mean skin temperature and a number of methods for

Table 3.1 Relative areas of body regions according to Hardy and DuBois[14].

Region	Relative area	Approximate integral proportions (c)
Head	0.07	1
Arms	0.14	2
Hands	0.05	1
Feet	0.07	1
Legs	0.13	2
Thighs	0.19	3
Trunk	0.35	5
Total	1.00	Total 15

determining this have been used. The techniques generally have been to measure skin temperature at a number of specific sites and to multiply each value by a 'weighting' factor equal to the fraction of the total body area at each temperature. Different weighting factors have been used according to the number of skin sites but 15 were proposed by Winslow *et al.*[13] and this has been found to be a reasonable maximum.

The most comprehensive method of estimating mean skin temperature by the use of a weighting system would be to use the 15 sites of Winslow and to calculate the arithmetic mean for each region and then weight these results for each subject according to the relative areas of the seven regions of the Hardy and DuBois formula[14] as shown in Table 3.1. Referring to Fig. 3.11 the mean temperature would then be expressed by the formula:

$$T_s = c_{head}(A+B+L)/3 + c_{arms}(D+F)/2 + c_{hands}(G) + c_{trunk}(C+E+M+N)/4$$
$$+ c_{thighs}(H+P)/2 + c_{legs}(J+Q)/2 + c_{feet}(K).$$

Such a formula is generally regarded as being too complicated and time consuming for routine use with many subjects and for this reason alternative expressions have been proposed, together with the idea that a single temperature at the medial thigh (position R in Figure 3.11) could give a fair approximation to mean skin temperature. In an analysis of the effect of environmental conditions on the agreement between different formulae Veghte[15] found

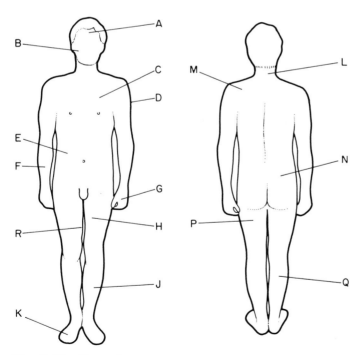

Fig. 3.11 Sites of spot temperature measurements used in the evaluation of mean skin temperature.

large differences when subjects were cold but good agreement when they were sweating. Moreover, Mitchell and Wyndham[16] examined a number of these weighting systems and compared them with an unweighted mean temperature from the readings of 15 sites distributed according to the relative areas of Hardy and Du Bois. This unweighted formula expressed mean skin temperature as follows:

$$T_s = (A + C + D + E + F + G + H + J + K + L + M + N + P + Q + R)/15.$$

They found that this last method was as reliable as any of the weighting formulae considered and was easier to use. The medial thigh temperature, although not at all equal to the mean skin temperature, was considered as being a useful indicator of, for instance, rate of change of mean skin temperature.

A novel method of measuring overall skin temperature was devised by Wolff[17] and made use of a vest constructed of electrical resistance wire. The vest could be worn next to the skin or between clothing layers, and the electrical resistance, measured with a galvanometer circuit, varied with the overall mean skin temperature.

Estimates of the mean skin temperature for regions and for the body as a whole can be made by analysing thermograms in terms of the proportion of the total area at each colour. The values obtained in this way can then be compared to those found from averages of probe measurements made at particular sites on the body surface. Such comparison was made during the investigations into temperature distribution over runners described earlier in this chapter (p. 52). The results showed (Figure 3.12) that regional temperature distributions were far from normal, in the statistical sense, and that the arithmetic mean values could differ by up to 4°C from spot readings taken with a thermocouple probe. In spite of these differences the overall arithmetic mean from 11–13 probe measurements was found to be within 1.5°C of the mean obtained from analysing the thermograms, thus agreeing with the findings of Mitchell and Wyndham.

In view of the complex nature of temperature distributions, it is not surprising that a small number of probe readings may give an inaccurate estimate of regional temperatures. The repeatability of such readings is also complicated by the need to measure at exactly the same position every time a measurement is taken. A positional error of a few centimetres can lead to temperature errors of several degrees. This was very clear on thermograms of the athletes where temperatures over the knees were at least 10°C lower than over the skin of the calf muscle, the distance between these sites being little more than 10 cm. One of the conclusions from these experiments was that it was unlikely, using probe techniques, that mean skin temperatures could be measured to better than ± 1.5°C in nude subjects at ambient temperatures of 20°C and below. A more complete analysis of infra-red thermography data and comparison with the traditional methods is desirable and is now a possibility with the introduction of digital methods for storing and analysing thermograms. The method of averaging up to 15 probe readings gives a good estimate of the mean skin temperature but indicates nothing about the particular distribution leading to the average.

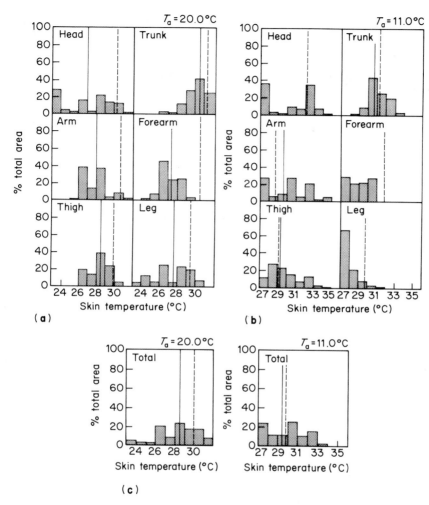

Fig. 3.12 Histograms obtained from thermographic analysis showing the area-temperature distributions in various body regions (a and b) and for the whole body (c) at two different environmental temperatures ($T_a = 20°C$ and $11°C$). The height of each column represents the percentage area occupied by each colour. Mean skin temperatures for each area obtained by thermography (————) are compared with those obtained by a thermocouple probe (- - - - - -). The absence of a thermographic mean in some regions is due to a significant percentage of the area being black and therefore lying outside the temperature range.[2]

A number of workers have found[18] that the central, or core, temperature varies with work rate and is independent of the environmental temperature whereas, on the other hand, mean skin temperature has been shown to be independent of work rate and to vary with the environmental temperature. In studies of the temperature distribution over runners, one of the most significant findings was the redistribution of temperature patterns during exercise when the warmest areas were coincident with the surface markings of the active muscles. Changes in skin blood flow, environmental cooling and evaporation completely redetermined the distribution. It is therefore perhaps fortuitous if mean skin surface temperature is independent of work rate and it is certainly true that the mean does not necessarily indicate any of the important changes that occur to the overall pattern of skin temperature distribution.

References

1.   Pugh, L.G.C.E., Corbett, J.L. and Johnson, R.H. (1967). Rectal temperatures, weight losses, and sweat rates in marathon running. *Journal of Applied Physiology* **23,** 347-52.

2.   Clark, R.P., Mullan, B.J. and Pugh, L.G.C.E. (1977). Skin temperatures during running—a study using infra-red colour thermography. *Journal of Physiology* **267,** 53-62.

3.   Clark, R.P. and Goff, M.R. (1979). Human skin temperature during rest and sleep visualized with colour infra-red thermography. *Journal of Physiology* **300,** 14-15P.

4.   Henane, R., Buguet, A., Roussel, B. and Bittel, J. (1977). Variations in evaporation and body temperatures during sleep in man. *Journal of Applied Physiology* **42,** 50-55.

5.   Rothwell, N.J. and Stock, M.J. (1979). A role for brown adipose tissue in diet-induced thermogenesis. *Nature* **281,** 31-5.

6.   Clark, R.P., Goff, M.R. and Garrow, J.S. (1983). Whole body quantitative infra-red thermography in obesity. *Proceedings of The International Conference on 'The Adipocyte and Obesity; cellular and molecular mechanisms'* June 1982. Toronto University, *International Journal of Obesity* **7,** No. 4, 403.

7.   Clark, R.P., Goff, M.R. and Mullan, B.J. (1977). Skin temperatures during sunbathing and some observations on the effect of hot and cold drinks on these temperatures. *Journal of Physiology* **267,** 8-9P.

8.   Clark, R.P. (1980). The effect of hot and cold drinks on skin surface temperature distribution. *Proceedings of the 9th International Colloquium on Coffee—London.* Vol 11, 405-11. Association Scientific Internationale Du Cafe (ASIC), Paris.

9.   Fox, R.H., Goldsmith, R. and Kidd, D.J. (1960). Cutaneous vasomotor control in human nose, lip and chin. *Journal of Physiology* **150,** 22-3P.

10.   Fox, R.H. and Hilton, S.M. (1958). Bradykinin formation in human skin as a factor in heat vasodilatation. *Journal of Physiology* **142,** 219-32.

11.   *Recent advances in Medical Thermology* (1984). Ed. E.F.J. Ring and B. Phillips. Plenum, New York.

12.   Ring, E.F.J. and Bacon, P.A. (1977). Quantitative thermographic assessment of inositol nicotinate therapy in Raynaud's syndrome. *Journal of International Medical Research* **5,** 217-22.

13.   Winslow, C-E.A., Herrington, L.P. and Gagge, A.P. (1937). Physiological reactions of the human body to varying environmental temperatures. *American Journal of Physiology* **120,** 1-14.

14.   Hardy, J.D. and DuBois, E.F. (1938). Basal metabolism, radiation, convection and evaporation at temperatures of 22/31°C. *Journal of Nutrition* **15,** 477-97.

15.   Veghte, J.H. (1965). *Infra-red thermography of subjects in diverse environments.*

Report AAL-TR-65-18, Arctic Aeromedical Laboratory, Aerospace Medical Division, USAF Systems Command.

16. Mitchell, D. and Wyndham, C.H. (1969). Comparison of weighting formulas for calculating mean skin temperature. *Journal of Applied Physiology* **26,** 622-8.

17. Wolff, H.S. (1958), A knitted wire fabric for measuring mean skin temperature and for body heating. *Journal of Physiology* **142,** 1-2P.

18. Kerslake, D. McK. (1972). The stress of hot environments. *Monograph of the Physiological Society*, Cambridge University Press, Cambridge.

19. Clark, R.P. and Goff, M.R. (1984). Infrared thermographic measurements of hand-skin temperature recovery following cold stress. *Journal of Physiology* **350,** 12P.

20. Clark, R.P. and Croft, M.R. (1983). The thermal cost of cutaneous temperature. *Journal of Physiology* **345,** 136P.

4
Convection

Convective heat loss may be considered in two distinct parts; there is free convection, which occurs in substantially 'still' air, and forced convection, where the body is either in an airstream or is moving through the air. Forced convection heat loss is often complicated by the fact that during movement the limbs do not move with uniform translation and that outdoor airstreams are rarely unidirectional or steady.

Free convection

The skin or clothing is generally at a higher temperature than the surroundings and this means that the air next to, and touching the body surface becomes heated by direct conduction; it then becomes less dense, more buoy-

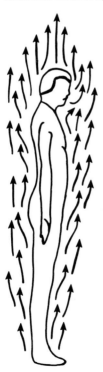

Fig. 4.1 Diagrammatic representation of the natural convective boundary layer flow generated around a standing subject. The air movement becomes faster and the layer thicker towards the head. This diagram may be compared with the Schlieren visualization of the flows seen in Plate 16(a).[22]

ant and, because of Archimedes' principle, begins to rise. The result of this happening over the whole of the body surface is to generate an upward moving envelope of warm air surrounding the body which increases in thickness and speed as it approaches the head. This flow of air is shown diagrammatically in Figure 4.1 and is termed 'the natural convection boundary layer'; it forms part of the human microenvironment and is the mechanism by which heat is carried away from the body surface by convection.

Visualization of the convective flow

Using special optical equipment, it is possible to visualize and photograph the natural convection boundary layer. The technique is known as Schlieren photography and had its origins in Germany at the end of the last century where it was used in the glass industry to detect flaws, or Schlieren, in high-quality plate glass. Schlieren photography is an established technique in the

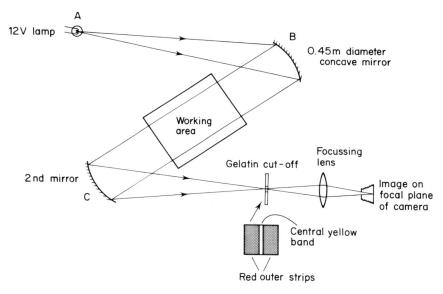

Fig. 4.2 Diagram of the Schlieren optical system used to visualize convective airflows in the human microenvironment. A light source (A), with a vertical filament is positioned at the principle focus of a 0.45 m diameter parabolic mirror (B). The light is reflected from this first mirror to travel some 7 m as a parallel beam to fill the 0.45 m diameter reflecting surface of a second mirror (C). From here the light is reflected to produce an image of the filament at the principle focus of the second mirror where a cut-off is arranged. This cut-off consists of a filter of three pieces of gelatine sheet, generally with yellow in the centre and red at the edges. Any change in the air temperature, and hence the refractive index of the air within the working area produce deflections in the beam focused at the edge of the cut-off. Modulation of the light beam in synchrony with thermal changes in the working area enables moving airstreams to be viewed as one colour against a background of a contrasting colour.[22]

aeronautical industry where it is used to visualize pressure and shock waves over aircraft models in high speed wind tunnels[1]. The recent use of the method to visualize airflows around the body has greatly helped our understanding of the physical and physiological conditions found in the human micro-environment[2, 3].

The Schlieren method relies on differences in refractive index developed in the boundary layer by the temperature gradients in the moving airstreams. Figure 4.2 shows the diagrammatic layout of the optical system currently in use at the Medical Research Council (MRC). The light source is a bulb with a vertical line filament positioned at the principle focus of a spherical mirror of high optical quality. Light is reflected from this first mirror as a parallel beam to travel some 7 m to the reflecting surface of a second mirror. Here, the light is reflected to form an image of the filament at the principle focus of the second mirror where a 'cut-off' consisting of a filter of coloured gelatin is arranged. Variations in the air temperature change the refractive index of the air within the working area to produce deflections of the image focused at the edge of the gelatin filter. The deflection or modulation of the focused light beam enables moving air to be seen as a coloured image against a background of a contrasting colour. Plate 16 shows Schlieren pictures of the flow visualized in this way over a standing subject.

The method has been used to study systematically the natural convective flow over the whole body surface and the following sections describe the features that have been observed using the 0.45 m diameter MRC system. The technique has a sensitivity that is dependent on the refractive index gradient within the flow and this means that generally about one-half of the total thickness of the boundary layer can be visualized.

Description of the flow over a naked standing subject

Starting at the feet, the major part of the moving air adjacent to the skin of the dorsa detaches itself and, because the surface is nearly horizontal, rises upwards away from the feet. Some of the warmed air, however, remains attached to the skin around the ankle where it joins with the air moving over the lower part of the leg. As the flow rises up the front of the leg the airspeed increases and the boundary layer becomes thicker. From here the layer of air, by now about 1–2 cm thick, passes the knee and thigh to the groin where the inner part of the layer continues to accelerate. The outer, and more turbulent part, becomes partially detached at the convexities and there appear to be regions of reversed flow. This pattern persists up to the chest level, until in the region of the shoulder, where the surface is again horizontal, most of the flow breaks away from the skin surface and continues upwards.

Despite the vigorous general upflow, there are regions where the air is brought to rest on the body surface; these areas are known as stagnation points and are found in the perineum, the axilla, under the lobe of the ear and beneath the nasal septum.

The airflow is modified by the contours of the neck, jaw and face. Part of the rising layer takes the line of least resistance and is deflected past the side of the head, while the remainder follows the undersurface of the chin. The pinna of the ear acts as a deflector for some of the rising warm air; the rest of the flow streams away from the side of the head.

The convective flow passes over the forehead to join the air which has been rising past the sides of the head and neck and produces a plume extending for more than 1 m above the head where the well-defined airstreams disintegrate and merge with the ambient air.

This account of the free convective airflow around the human body is different from that often given in previous books on temperature regulation in man and animals; these usually describe a layer of air of uniform thickness surrounding the body and insulating it against heat loss. The thickness of such a layer was considered to be reduced by air movement to account for the fact that airstreams produced an increased body cooling.

The new description of free convection given here emphasizes several points; first that this layer of warmed air is not static but is moving at up to 0.5 m/s over the body surface. Secondly, the moving air provides an insulation but one which is not uniform; the flow increases in thickness as it passes upwards over the body surface, so that by the time it reaches the head and neck and especially the face, it is many centimetres thick and here provides its maximum insulation. This is one reason why our naked faces do not feel cool or cold except in cold weather with a strong wind blowing. In quite cool rooms the face may feel perfectly comfortable because it has this thick layer of warm air surrounding it. A further point is that this airflow provides us with our own 'air-conditioning system' for the head, neck and face which prevents a build-up of expired CO_2, this is particularly important for infants nursed in cots or incubators and when adults lie down or recline, in providing a fresh supply of air, free from CO_2 for inspiration.

Mathematical description of the natural convective flow

A mathematical analysis can be made of the convective flow and the heat output from the human body, using boundary layer theory. There is a non-dimensional parameter that characterizes the natural convection flow known as the Grashof number; this is defined by:

$$Gr = \frac{gh^3}{v^2} \frac{(T_s - T_a)}{T_a}$$

where g is the acceleration due to gravity, v is the kinematic viscosity of the air, h is the vertical height on the body surface, T_s is the skin temperature and T_a is the air temperature.

The Grashof number is proportional to the ratio of the buoyancy to the viscous forces within the airflow; when it is less than 10^9 the boundary layer is described as being laminar and it becomes fully turbulent when the Grashof number exceeds 10^{10}.

For a naked standing man, where there is a temperature difference of 6–10°C between the skin and the surroundings, the boundary layer is laminar up to a height of approximately 1 m (about the level of the navel) and is fully turbulent after some 1.5 m as illustrated in Figure 4.3.

An examination of the mathematical expression for the Grashof number

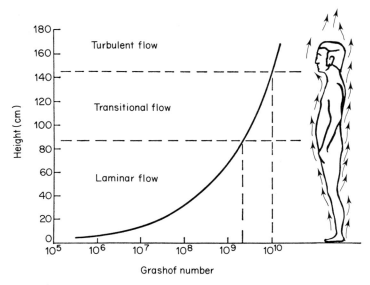

Fig. 4.3 Variation of Grashof number with vertical height over the body surface for a mean skin temperature of 33°C and an ambient temperature of 25°C.[13]

shows the temperature difference, between the skin or clothing and the surrounding air, appearing to the single power. This means that if the temperature difference doubles there is only a twofold change in the Grashof number. On the other hand, the vertical height appears as h^3 in the expression and a doubling of h results in an eightfold (nearly one order of magnitude) change in the Grashof number. Thus the nature of the flow, whether it is laminar or turbulent, is highly dependant on the height of the person and is less sensitive to the temperature difference between the skin or clothes and the surrounding air.

The Schlieren technique can demonstrate that clothing modifies the convective pattern, and since there can be a surface temperature difference of several degrees due to the insulation produced by clothing it is possible for the boundary layer over the whole surface of a standing, clothed person to remain laminar, that is, the Grashof number will not exceed 10^9.

Air velocities and temperature profiles found within the boundary layer

Large quantities of air move in the human microenvironment and at surprisingly high velocities. For the naked standing man having an overall mean skin temperature of 33°C in a room at 25°C, the natural convective boundary layer has a thickness of about 180 mm at the level of the face. The maximum velocity found at this height is about 0.5 m/s and occurs some 20 mm away from the skin surface. The total quantity of air passing over the head is about 10 litres/s and the fast-moving plume of air extends for 1–1.5 m above the head.

A general appreciation of the velocity and approximate total extent of the boundary layer may be established from the Schlieren visualization. However, the system is not suitable for more detailed quantitative assessment of the flow and in order to measure velocity and temperature profiles within the boundary layer, thermocouples and anemometer probes have to be used.

In the original work to establish the boundary layer parameters, heated models were constructed to help develop measuring techniques sensitive enough for use in the natural convective airflow[4]. Figure 4.4 shows these models, which produced a convective boundary layer similar to that found over man but with the advantage that they did not have the small involuntary movements which make measurements difficult over human subjects. Figures 4.5(a) and 4.5(b) are typical temperature and velocity profiles measured in both the laminar and turbulent parts of the boundary layer flow.

The shape of the velocity profile in the human convective flow is such that there is zero velocity at the skin surface (the skin does not move, so neither

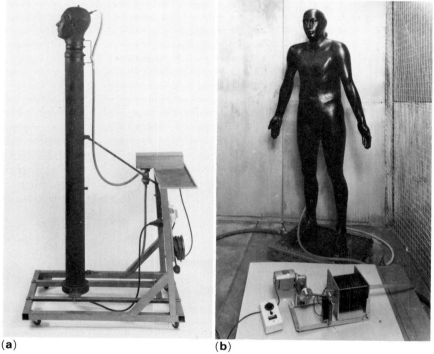

(a)   (b)

Fig. 4.4 Two manikins used to investigate aspects of the natural convective airflows found in the human microenvironment. (a) A cylindrical heated model producing a thermal boundary layer with known characteristics and having velocity and temperature profiles similar to those found over human subjects. (b) A full-sized heated model of the human form used to investigate the effects of non-geometrical shape on the convective flow pattern. In addition, this manikin had an artificial 'breathing' system to study the interaction of respiratory flows with the thermal boundary layer.

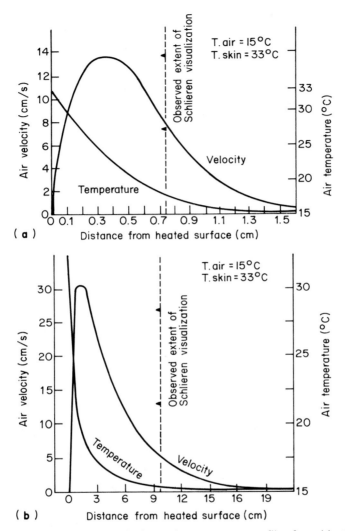

Fig. 4.5 Typical velocity and temperature profiles found in the laminar (a) and turbulent (b) regions of the human microenvironment. Velocity reaches a maximum at about one-third of the total boundary layer thickness from the skin surface. Schlieren visualization of the flows is inherently limited to about half the total flow thickness.[13]

can the air that is actually in contact with it). Airspeed increases sharply with distance from the skin surface to reach a maximum about one-third of the way across the boundary layer; the velocity then begins to fall to reach zero again towards the outer part of the layer. Figure 4.6 shows the shape of the velocity profile and how the flow develops along the front part of the legs between the ankle and knee with the velocity and thickness steadily increasing.

Scale of ordinates 0 10 20 30 cm/s

Scale of abscissa   0 1 2 3 cm

Scale of leg   0 2 4 6 cm

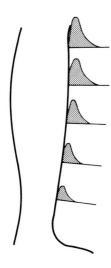

Fig. 4.6 Diagram showing the development of the velocity profile up the front surface of the leg. These curves were obtained on the heated manikin seen in Fig. 4.4(b) from a hot wire anemometer traverse.[13]

Heat loss in the boundary layer flow

The heat transfer by natural convection in the boundary layer flow is determined by the slope of the temperature profile and can be described by the equation kdT/dY where k is the thermal conductivity of the air and dT/dY is the horizontal temperature gradient in the airflow at the body surface. This expression shows that the heat loss is directly proportional to the temperature gradient, that is the steeper the fall in temperature, the higher will be the heat loss. Conversely if the temperature gradient is shallow there will be less convective heat loss.

If we consider the temperature profile in the boundary layer we can easily see how the flow thickness can have a direct influence on the heat loss. Where the boundary layer is thin, at the ankles and lower leg, the temperature gradient is steep and the temperature has to fall from the value at the skin surface to that of the environment within the thickness of the boundary layer. On the other hand, where the boundary layer is thicker, over the head and upper parts of the body, the temperature has a greater distance in which to reach the ambient value; the gradient is shallower and consequently the convective heat loss less. In summary, the thicker the boundary layer, the lower the convective heat loss.

This may be illustrated by the flow pattern and temperature and heat loss distribution around a heated hollow sphere. Figure 4.7(a) shows the convective flow patterns around a 13 cm diameter sphere when viewed in the Schlieren optical system. The convective flow is thin at the bottom but thickens around the circumference towards the top when the airstreams form an upward moving hot 'plume'. When the temperature and local heat transfer coefficients (of convection and radiation measured using the heat transfer calorimeter described in Chapter 2) are measured around the circumference it is found that where the flow is thinnest, and the temperature gradient steepest, the surface temperature is lowest but the heat transfer highest (Figures 4.7(a) and 4.7(b)). As the flow thickens the temperature gradients in

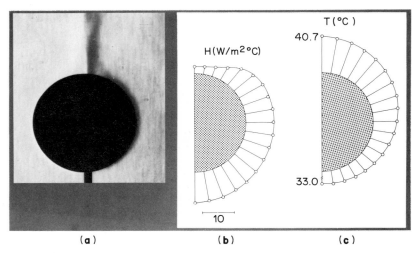

(a)   **(b)**   **(c)**

Fig. 4.7 An illustration of the relationship between convective flow patterns, local heat loss and surface temperature for a heated hollow sphere. (a) shows the convective flow patterns visualized with the Schlieren technique with the thin boundary layer and slow moving air at the underside of the sphere. The layer becomes thicker and faster around the circumference to eventually produce a plume of warm air streaming away from the top surface. (b) Convective and radiative heat loss patterns plotted around the circumference (radial distance from the circumference being proportional to heat loss) showing greatest heat loss from the lower surfaces of the sphere where the convective flow is thinnest and slowest. (c) Surface temperature on the sphere showing highest values at regions of lowest heat loss and lowest temperatures where the heat loss is greatest.

the airstreams become smaller and the heat loss reduces, but surface temperature rises because in this case the sphere is hollow. At the top of the sphere, in the 'plume' where the flow is thickest the temperature is highest and the heat transfer lowest. This principle, relating boundary layer thickness to convective heat loss applies to all parts of the human body and enables flow patterns and heat exchange rates over various areas and in different postures to be described. It is also important in understanding the terms 'hot' and 'cold' as discussed in Chapter 1. The bottom of the sphere has the lowest temperature as a consequence of the high heat loss and on this basis could be considered to be hot. On the other hand the heat loss at the top of the cylinder is lowest but the temperature highest and so the top could be called the hottest. Great care must be taken in the use of the terms 'hot' and 'cold' where the meaning is not precisely defined.

Boundary layer equations for heat loss

Mathematical expressions for the velocity and temperature profiles found within the natural convection boundary layer can be combined in terms of the physical properties of the air, and the geometrical parameters of the body surface to describe the basic relation that heat loss is directly proportional to the temperature gradient at the surface[5, 6]. Because of the complexity of the fluid dynamics of these phenomena they can only be evaluated for specific shaped bodies and if the equations are examined for a vertical plate of height h and width b the expression for convective heat loss becomes:

$$Q = \frac{0.34bh^{\frac{3}{4}}k}{v} \sqrt{g\frac{(T_s - T_a)^3}{T_a}} \text{ watts}$$

In order to apply this analysis to evaluate human convective heat loss, it is convenient to consider the body as made up of a series of cylinders representing the head, torso, the arms and legs. Figure 4.8(a) illustrates the dimensions of such a set of cylinders which may be further simplified by opening the cylinders to form flat plates (Fig. 4.8(b)) and by considering the torso and legs together as one; this final set of plates and their dimensions is shown in Figure 4.8(c). Table 4.1 shows the convective heat output in watts at three

Table 4.1   Convective heat output from the plates shown in Fig. 4.9(b).

Body region	Heat output (watts) at environmental temperature of			Percentage of total
	15°C	20°C	25°C	
Each arm	12.9	7.9	3.8	13.8
Trunk and legs	57.0	35.0	16.9	61.0
Head	10.8	6.6	3.2	11.4

different ambient temperatures for the plates when the above heat transfer equation is used. Plate surface temperature has been taken as 33°C to represent the mean value for skin temperature.

These calculations provide a simplified estimate of heat loss without taking account of body features such as the hair on the head or the relatively large surface area of, for example, the face or fingers; such details are at present beyond the scope of existing theory.

There have been a number of attempts[7] to measure free convective heat transfer coefficients directly from the human body but it is only with the

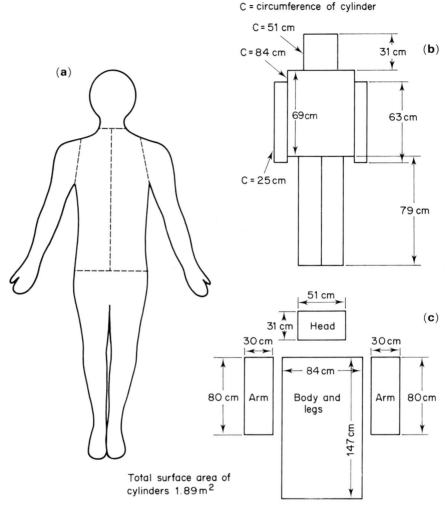

Fig. 4.8 Typical example of the division of the human body (a) into a series of cylinders (b) and reduction to flat plates (c) for heat transfer analysis.

advent of techniques such as the naphthalene sublimation method of Nishi and Gagge[8] and the surface calorimetric method developed by Toy (described in Chapter 2) that accurate direct assessments of convective heat loss are possible.

Figure 4.9 shows free convection heat loss rate determined indirectly for man, plotted against skin (or clothing) to air temperature difference. Also shown is an estimate based on Birkebak's[11] simplified equation for a vertical cylinder 2 m high. Table 4.2 gives free convective coefficients for vertical and horizontal plates/cylinders in air also compiled from Birkebak[11].

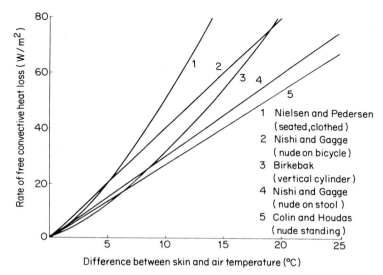

Fig. 4.9 Rates of free convective heat transfer for man according to Nielsen and Pedersen (1952)[9], Colin and Houdas (1967)[10], and Nishi and Gagge (1970)[8] compared to the rate for a vertical cylinder calculated from Birkebak (1966)[11]. Reproduced with permission from Mitchell (1974)[12]. *Physical Basis of Thermoregulation.* International Review of Science and Physiology Series, Vol. 7. Environmental Physiology. Ed by D. Robertshaw. Medical and Technical Publishing Co., Lancaster.

Table 4.2 Simplified free convective heat transfer coefficients (h_c) for vertical plates or cylinders and horizontal cylinders in air (modified from Birkebak[11]).

	Characteristic dimension L (m)	Applicable range of $L^3 \triangle T$	h_c (W m^{-2} °C^{-1})
Vertical plates or cylinders	Height	1.5×10^{-4} to 15 15 to 1.5×10^{-4}	$1.4(T/L)^{\frac{1}{4}}$ $1.5\ T^{\frac{1}{3}}$
Horizontal cylinders	Diameter	1.5×10^{-4} to 15 15 to 1.5×10^4	$1.2\ (T/L)^{\frac{1}{4}}$ $1.2\ T^{\frac{1}{3}}$

The average value for the convective cooling coefficient in the example given in Table 4.1 is 2.3 W/m².°C and between 2.7–3.0 W/m².°C from the results in Table 4.2. It is encouraging that these various estimates seem in reasonable agreement at moderate values of temperature difference and produce a value of 30–40 W/m² of convective heat loss from the body for a 10°C temperature difference.

Effect of posture on natural convective heat loss

Once it is appreciated how the thickness of the convective boundary layer and the temperature gradients directly determine the heat transfer it is readily seen that changes in posture will alter the flow patterns and consequently modify the heat output. Schlieren photography has enabled the convective airstreams over the body in various postures to be visualized. Figure 4.10 illustrates diagrammatically the convective flow patterns for the lying, sitting and standing postures.

Fig. 4.10 Diagrammatic representation of the natural convective boundary layer flow generated over a subject in the standing, sitting and lying postures. When lying, the flows are generally slower and thinner than when standing. Sitting produces flows which are intermediate between lying and standing.

Postural changes produce the largest differences in flow patterns over the face and head and the resultant heat loss has been examined in some detail[13, 14]. Figure 4.11 illustrates the flow patterns over the head; for the standing man, the flow over the face is that which has been developing in velocity and thickness over the whole height of the body. This means that the boundary layer is thick, the temperature gradient shallow, and consequently the convective heat loss is low.

In contrast, the flow over the head and face of the subject in the lying position is quite different. The convective flow is only that which has developed over the head itself and the flow over the lower part of the body has no influence when the subject is lying down. The maximum air velocities found over the face in the supine posture are only some 0.04–0.05 m/s compared with 0.3–0.5 m/s when standing. Thus the insulating properties of a thick boundary layer are effectively lost when lying down.

With a seated subject the flow pattern is more complicated; a boundary layer begins to form from the lower abdomen and interacts with the airstreams

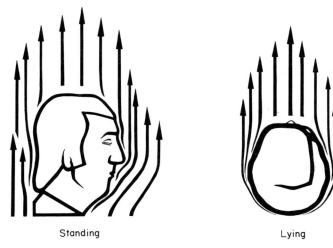

Standing   Lying

Fig. 4.11   Diagram of the convective flow patterns over the head and face when standing and lying. When lying, the flows are slow moving and thin, unlike in the standing posture when they have been developing in speed and thickness over the lower part of the body.[22]

that break away from the horizontal surfaces of the knees and thighs. The result is that the flow over the face differs little from that found in the standing posture.

As we have seen, the convective heat loss from the skin is related to the exact nature of the natural convective flow patterns and this can be illustrated by considering the results of heat loss measurements from different parts of the face. Figure 4.12 shows local convective heat loss values at five sites over the head of a subject when supine and prone. Comparison with Figure 4.7 shows that the heat loss pattern is similar to that around the cylinder and is therefore directly related to the thickness of the convective airstreams. At upward facing parts the air flows are thickest and the temperature gradients small and therefore the convective heat loss low.

The local convective measurements show the detailed relationship between heat loss and flow thickness, but in order to assess the total heat loss from the head in any posture it is necessary to make a number of measurements at several sites and to average the results. Experiments where this has been done have shown that the heat loss from the head when lying down is some 30 per cent higher than when standing. These results have provided the most detailed information so far available about the interrelation between heat transfer and natural convective air flows in the human microenvironment. Also the techniques that were developed for this work are at present the only means available of quantifying the considerable heat transfer changes associated with posture or other modifications to the convective flow pattern. The analysis of heat transfer equations leading to the simplified coefficients shown in Table 4.2 are not sensitive enough to reveal such phenomena. This can be illustrated if we use the coefficients to determine the total heat loss from a

Heat loss (W/m^2. °C)

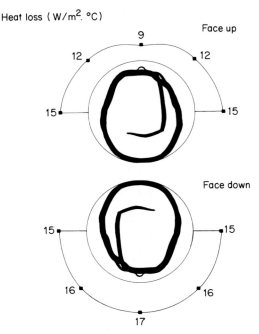

Fig. 4.12   Local convective and radiative coefficients (W/m².°C) from the face for a supine and prone subject with a full head of hair measured with surface plate calorimeters. Mean skin temperature 33°C and ambient temperature 23°C. (Reproduced with permission from the *Journal of Physiology*)[13].

cylinder 2.0 m high, 0.29 m diameter (surface area 1.8 m²) with a surface temperature of 33°C in an environment of 25°C ($\triangle T = 8$°C). The results of such calculations are shown in Table 4.3 where the heat loss from the vertical cylinder is approximately 12 per cent more than for the horizontal one. Analysis such as this is quite inappropriate to demonstrate the dependence of

Table 4.3   Heat loss from vertical and horizontal cylinders.

	Expression for convective coeff. h_c	h_c (W/m². °C)	Heat output (watts)
Vertical cylinder	$1.5(\triangle T)$	3.00	43.2
Horizontal cylinder	$1.2[(\triangle T)/L]^{\frac{1}{4}}$	2.69	38.7

$\triangle T$ = cylinder—environmental temperature difference
  L = 2 m (height) for vertical cylinder, 0.29 m (diameter) for horizontal cylinder.

local heat loss on microenvironmental features, and indeed this illustration shows that there are no very large differences in total heat loss when lying down or standing up. Nevertheless, such overall simplified analyses are still useful in assessing the order of magnitude of total heat loss from the body surface.

Forced convection flow patterns

When the body is exposed to a wind or when it is in movement through the air, the natural convective airstreams are displaced and heat is then lost by forced convection. It is again convenient to represent the body as a heated cylinder in order to illustrate some of the physics of the forced convective process. When the unheated cylinder is in a unidirectional flow, the air at the front leading edge, known as the stagnation region, is brought to rest, whilst adjacent air is forced to speed up and pass around the side of the cylinder. About half way around the circumference the inertia of the air next to the surface becomes too great for the flow to adhere any longer, and it breaks away to form a wake.

The characteristics of the boundary layer and airflow are determined by the speed and viscosity of the air and the diameter of the cylinder. These parameters can be arranged in a non-dimensional group known as the Reynolds number which is defined by:

$$R = \frac{\text{Air velocity (V)} \times \text{characteristic body dimension (D)}}{\text{Kinematic air viscosity } (v)}$$

The value of the Reynolds number determines whether the forced airflow (as opposed to convective flow where the Grashof number determines the characteristics) is laminar or turbulent; in laminar flow, when the Reynolds number is less than about 20 000, the air breaks away from the cylinder

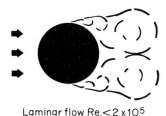

Laminar flow Re.$<2 \times 10^5$

Turbulent flow Re.$>2 \times 10^5$

Fig. 4.13 Flow separation from the surface of a cylinder in laminar and turbulent flow. The Reynolds number (Vd/v) (where V is the air velocity, d the cylinder diameter and v the kinematic viscosity of the air) determines the type of flow and the point at which it breaks away from the surface to leave a wide wake in laminar flow or a narrow wake when turbulence is reached. (Reproduced with permission from the *Journal of Physiology*)[13].

surface at about 90° from the leading edge. If the Reynolds number is greater than this the flow is fully turbulent and breaks away further around the cylinder to form a narrower wake. Figure 4.13 shows diagrammatically this effect for both laminar and turbulent flow.

Heat loss in forced convection

For the case of a heated cylinder in a moving airstream the resultant flow patterns are determined by the interaction of the free convective flows pre-

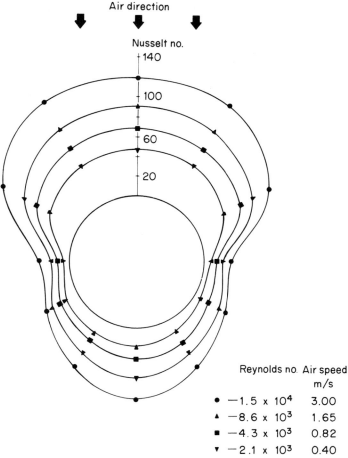

	Reynolds no.	Air speed m/s
●	−1.5 x 10⁴	3.00
▲	−8.6 x 10³	1.65
■	−4.3 x 10³	0.82
▼	−2.1 x 10³	0.40

Fig. 4.14   Diagram showing forced convective heat transfer distribution patterns (expressed as Nusselt numbers i.e. $h_c d/k$ where h_c is the local heat transfer co-efficient, d the cylinder diameter and k the thermal conductivity of the air) for different airspeeds and Reynolds numbers. The greatest heat loss occurs at parts of the cylinder facing into the airstream, and the lowest loss is at the sides where the flow separates. Further around, in the wake, the heat loss again increases. (Reproduced with permission from the *Journal of Physiology*)[16].

viously described, which are dependent on the Grashof number, and those caused by the moving air and characterized by the Reynolds number. At this stage it is convenient to introduce two further non-dimensional mathematical parameters. The first is the Nusselt number defined as $h_c L/k$ where h_c is the convective heat loss coefficient (W/m². °C), L a characteristic dimension and k the thermal conductivity of the medium (air or water) surrounding the body.

The other non-dimensional term is the Prandtl number, Pr, which is expressed as $\mu Cp/k$ where μ is the dynamic viscosity and Cp the specific heat (at constant pressure) of the fluid medium. The Prandtl number can relate momentum transfer expressed in terms of the Reynolds number, to the heat transfer described using the Nusselt number (i.e. $Nu = f(Re, Pr)$). The Prandtl number is composed only of the physical properties of the fluid and is not dependent upon air patterns or velocity and it is useful for comparison of heat transfer rates in air and water where Pr has the value of 0.71 and 7 respectively (at approximately 20°C). The Reynolds number is determined by the fluid flow pattern and for similarly shaped bodies in the same fluid, equality of Re implies similar values for the Nusselt number.

Fluid dynamic analysis of the resultant flow patterns and associated heat loss coefficients is complex and only the results that are of physiological

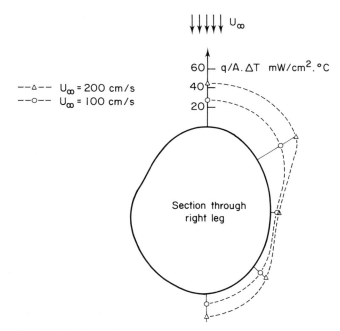

Fig. 4.15 Forced convective heat transfer profiles around a leg exposed to an airstream measured with surface plate calorimeters. The heat loss pattern has features consistent with those found around a heated vertical cylinder in an airstream.

significance will be mentioned here. We have already seen that in natural convection, the local heat loss depends on the value of dT/dY, the temperature gradient at the skin surface. Although the mathematical analysis is more involved the same principle holds for forced convection and the resulting convective heat loss distributions around heated cylinders in airstreams are well established both theoretically and experimentally, and are illustrated in Figure 4.14. At the front leading edge, where the flow is brought to rest, the temperature and velocity gradients are steep, causing a high local heat loss. At the side, where the flow breaks away, the gradients are less due to flow reversals in the separation region, and the heat loss reaches a minimum value. Further around, in the wake, the temperature gradients increase and heat loss rises.

A graphical integration of the distribution gives an average value for the heat loss for the whole cylinder, enabling an overall heat transfer coefficient to be calculated.

In many respects the heat loss pattern around the heated cylinder is similar to that over the human body. For instance, when the convective heat loss around the leg of a subject standing in a wind tunnel at airspeeds of 1 and 2 m/s is measured[15] the pattern shown in Figure 4.15 results. At the front leading edge the heat loss is highest and where the flow breaks away at the side, it reaches a minimum value; further around, in the wake, the heat loss again rises. The airspeeds of 1 and 2 m/s represent values of Reynolds number between 1.4×10^4 and 2.7×10^4 which is just within the range where the flow is still laminar.

Further evidence of the dependence of heat loss distribution on airflow pattern is seen in measurements obtained from the human head in a uniform airstream[16]. Figure 4.16 shows that this distribution is similar in nature to that found over the heated cylinder. These results were obtained from measurements over the median plane of the head of a bald subject with air velocities ranging from 0.15 to 1.4 m/s. At the higher airspeeds the expected pattern is seen with the highest heat loss at the front and the lowest, in this case at the top of the head, where the flow breaks away from the skin surface. The lowest of the horizontal air velocities of 0.15 m/s was lower than the maximum found in the natural convection airstreams in still air, and this resulted in a mixture of free and forced convective flow. The heat loss at this low velocity differed from the true forced convective pattern by being fairly constant around the face but lower on the top of the head where the convective plume had not been fully displaced by the horizontal airstreams.

A number of workers have measured forced convective cooling coefficients directly and indirectly from subjects exposed to uniform airstreams. If overall convective heat transfer coefficients for sections of the body, or the body as a whole, are plotted against the square root of the mean air velocity a substantially straight line results. Figure 4.17 shows this relation obtained from the experimental results of several workers. Kerslake[17] has summarized this work and suggests that for practical purposes, the following mathematical expression relating convective heat loss (h_c) to airspeed is

$$h_c = 8.3 \sqrt{V} (W/m^2 \cdot {}^\circ C)$$

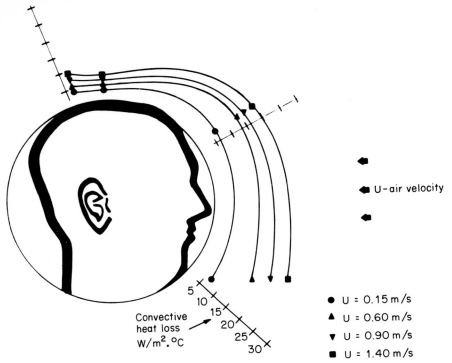

U – air velocity

● U = 0.15 m/s
▲ U = 0.60 m/s
▼ U = 0.90 m/s
■ U = 1.40 m/s

Convective
heat loss
W/m².°C

5
10
15
20
25
30

Fig. 4.16 Local convective heat loss coefficients, h_c (W/m².°C), on the head of a bald subject in forced convection with air velocities ranging between 0.15 and 1.4 m/s. The highest heat loss occurs at the front 'leading' surfaces and the lowest is at the top of the head where the flow breaks away. (Reproduced with permission from the *Journal of Physiology*.)[16]

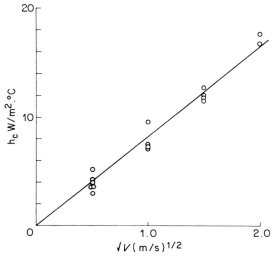

h_c W/m².°C

20

10

0

1.0

2.0

$\sqrt{V}$ (m/s)$^{1/2}$

Fig. 4.17 Relationship between convection coefficient, h_c, and the square root of wind velocity. Reproduced with permission from Kerslake (1972)[17].

Physiological aspects of forced convection

As we have seen, there are many situations where the body heat loss is similar to that from heated models. However, this analogy cannot be universally used to predict body heat transfer as, unlike the most sophisticated model, the body can alter heat loss by modifying surface temperature, by varying the blood supply to the skin and also by changing sweat rate and evaporation. This can have several results; one is to modify the actual quantity of heat available at the skin surface for dissipation to the environment, and the other is to change the skin and tissue conductance and thereby alter the amount of heat transmitted through the tissues from the central warm regions.

Areas over the head and face in particular, are sites where such effects may be demonstrated. Figure 4.18(a) and 4.18(b) show the variation of skin

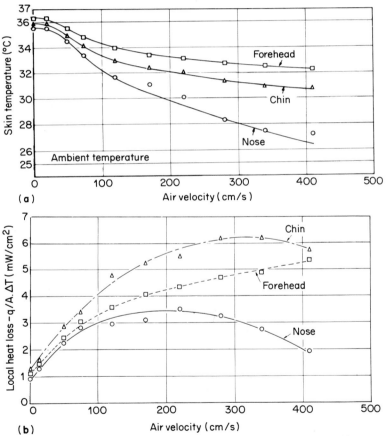

Fig. 4.18 Relationship between local skin temperature (a) and heat transfer (b) (measured with surface plate calorimeters) for three sites on the face of a subject in wind at different speeds. Temperature steadily decreases with increasing wind speed whereas heat loss increases to reach a maximum for the nose and chin at between 2 and 3.5 m/s.

temperature and heat loss during exposure to increasing wind speed over the forehead, chin and nose. The temperatures of these areas steadily dropped as wind speed was increased to 4 m/s. The rate of temperature drop was highest over the nose where the final temperature was only 1.5°C higher than the environment.

The local heat loss measurements show that the forehead steadily lost heat, whereas the chin and nose had a heat loss which reached a maximum value and then began to decrease even though the wind speed was increased. This is in marked contrast to the heat loss from a uniformly heated model, having a constant quantity of heat available for dissipation, where the heat loss continues to rise with increasing airspeed. Such results illustrate the way in which the body may conserve heat by selective shutting down of the heat loss mechanisms.

The 'pendulum' effect

We have now seen that convective heat loss from the body in 'still' air and when exposed to a uniform airstream can be described theoretically and measured with reasonable accuracy. However, in practical environments, the body is rarely in uniform airstreams; out of doors the wind is turbulent and gusty and hardly ever steady. Indoors, air conditioning systems and body movements produce turbulent conditions and even in 'still' air when a person walks or runs, only a small part of the body moves in a linear or unidirectional fashion.

If we consider various parts of the body during movement, we find that only the head and trunk can be said to move with anything like steady translation through the air. The upper arms and thighs perform a 'pendulum' or swinging motion and the forearms and lower legs have a whiplash movement[18]. As may be expected, such movements produce different amounts of convective cooling than from parts of the body with a more uniform motion. When trying to assess heat loss from all of the moving parts of the body and also taking into account the complication of a turbulent wind if the subject is out doors, the overall heat transfer becomes extremely difficult to evaluate. The Schlieren method has helped to clarify some parts of this problem and observation of the airflows around the moving legs of a runner have shown that the 'pendulum' effect produces quite different airflow patterns from those found in linear airstreams or in uniform translation. Around the swinging thigh the front 'bow wave' and trailing wake are alternately established and reversed after every change in direction of the swinging leg. The flows around the lower legs and forearms, which perform a whiplash movement, are similar in nature although more complicated in detail.

Classical fluid dynamic heat transfer analysis is inappropriate for these situations, and the movements of the body during walking and running are far more complicated than those associated with man-made engineering structures where the analyses and formulae have traditionally been required.

In the case of the moving limbs of a runner, some heat loss measurements have been made to complement the Schlieren visualization. Figure 4.19 shows the sequential positions of the thigh and lower leg of an athlete running at

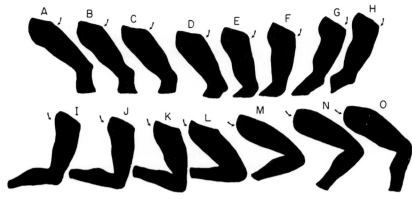

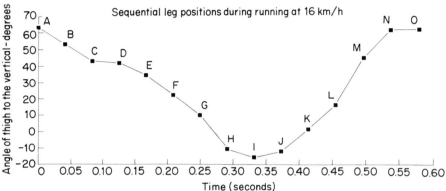

Fig. 4.19 Sequential angular positions of the thigh and lower leg of an athlete running at 16 km/h on a treadmill determined from cine film analysis.

16 km/h (10 m.p.h.) on a treadmill taken from a frame analysis of a cine film. The motion of the thigh relative to the body is roughly sinusoidal with a period of approximately 0.6 seconds. Heat transfer gauges attached to the thigh enabled measurements of temperature and heat loss to be made during running and Figure 4.20(a) and 4.20(b) show distributions around the thigh during standing and running both with and without wind (at running speed). The mean value for convective and radiative heat loss during standing was found to be 6.5 W/m². °C; during running at 4.5 m/s (10 m.p.h.) on a treadmill in 'still' air this rose to a mean value of 22 W/m². °C and to 54 W/m². °C whilst running in the presence of wind. The increase in heat loss during running in 'still' air was accompanied by a fall in skin temperature presumably brought about by environmental cooling due to the swinging and whiplash movements of the legs. The important finding is that the heat loss is very much higher than would occur in a uniform linear airstream and which could be evaluated using the formula

$$h_c = 8.3 \sqrt{V} \, (W/m^2 . °C)$$

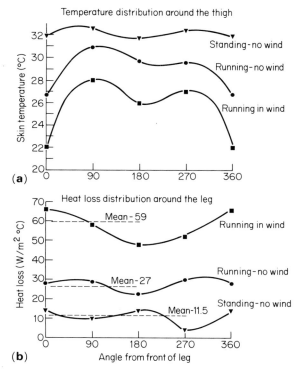

Fig. 4.20 Skin temperature (a) and local heat loss distribution (b) patterns around the thigh of the athlete during standing in still air, running (16 km/h) in still air and running in the presence of a 16 km/h wind.[22]

The swinging motion of the limbs therefore seems capable of producing a greatly increased heat loss from moving parts of the body and can explain how a gentle swinging of the arms on a hot day is so effective at cooling.

Previous studies[19] on the athlete taking part in these 'pendulum' experiments whilst running out of doors produced a convective cooling coefficient for the whole body of about 30 W/m². °C, compared to the value of 17.6 W/m². °C that would be expected at his running speed from the unidirectional airflow formula.

Detailed and comprehensive measurements of convective heat loss over the whole body in this complicated situation are not yet available but from the results so far it is likely that the high coefficients over all of the swinging parts of the body in motion through the air would be adequate to increase the average total convective cooling coefficient by a factor of approximately 2, compared with the uniform air movement situation.

The reason for the increased convective cooling around the swinging limbs is to be found by examining the circumferential convective cooling diagram. In conditions of true forced convection the circumferential pattern has large differences between the heat loss at the front leading edge and at the sides, in the wake of the cylinder as shown in Figure 4.14. The measurements around the swinging thigh of the runner produce a distribution that is more even around the circumference and it is the mean value for the distribution which determines the average cooling coefficient. These measurements confirm the visualization of the airflows using the Schlieren method which show that the alternating flow patterns are quite different from those seen in true forced convection. Undoubtedly, the 'pendulum' effect accounts for high, and previously unexplained, heat loss rates measured in exercising subjects by correlation of oxygen uptake with energy expenditure.

The reversed 'pendulum' effect

The study of the 'pendulum' effect in relation to the moving limbs and translation through the air has led to a consideration of heat loss in situations that could be described as being opposite to this, for instance when a subject is stationary in a turbulent or buffeting wind. In this case, if the scale of turbulence is large in relation to body size, the air will appear to the subject to be continually changing direction. Parts of the body facing directly into the airstream will, a short time after, be in a trailing wake and the opposite body area will face the airstream.

By analogy with the mechanism described in the previous section, this situation could be termed the reversed 'pendulum' effect. Such conditions have recently been studied in the harsh environment found beneath a hovering helicopter[20, 21] during some work which formed part of a trial to evaluate clothing suitable for personnel who work beneath helicopters on ships at sea in arctic conditions. Measurements made beneath these helicopters confirmed that the highly turbulent nature of the airflow produced, for a given mean windspeed, much higher cooling coefficients than would have been expected. Table 4.4 shows these results together with coefficients predicted from the formula for linear unidirectional airstreams.

Table 4.4 Cooling coefficients in highly turbulent airflow (beneath a hovering helicopter) compared with predicted coefficients.

Mean air speed m/s	Measured h_c (W/m².°C)	Predicted h_c (from $h_c = 8.3 \sqrt{V}$) W/m².°C
5	43	19
13	72	30
20	80	37

These results were obtained from a heated man-sized cylinder placed in the airflows beneath the helicopter. The cylinder was equipped with electrical heaters and surface temperature probes so that the quantity of heat required to maintain a given surface temperature could be measured in terms of the electrical power supplied to the heaters. Experiments were also performed to measure the heat loss when the cylinder was placed in a surface wind of

7 m/s (14 knots) having a much lower turbulence level and values of h_c of 30 W/m² . °C were obtained which were substantially higher than the 22 W/m² . °C expected from the linear airflow formula.

This section has attempted to outline some of the problems of measuring forced convective cooling coefficients from the human body in real situations, as summarized in Figure 4.21. The considerable amount of work carried out in uniform airstreams for 'still' subjects or models is only applicable to fairly limited conditions. The new approaches described here based upon the Schlieren technique and a method for measuring local heat transfer provide a foundation for future research that needs to be undertaken to establish comprehensive data on body heat transfer coefficients during movement and in non-linear airstreams.

Forced convective situations

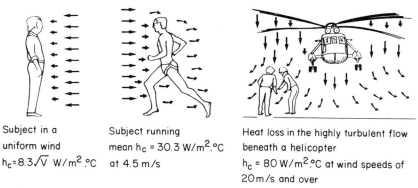

Subject in a uniform wind $h_c = 8.3\sqrt{V}$ W/m² . °C	Subject running mean $h_c = 30.3$ W/m². °C at 4.5 m/s	Heat loss in the highly turbulent flow beneath a helicopter $h_c = 80$ W/m². °C at wind speeds of 20 m/s and over

Fig. 4.21 Examples of forced convective heat loss in uniform and turbulent winds.

A further application of these principles is to be seen in the design of air-conditioning systems in the 'built environment'. Many such designs perform disastrously, mainly because they are specified by architects and constructed by air-conditioning engineers who have little experience of the interaction of their designs with the physiology of temperature regulation that can produce conditions of comfort or discomfort. Moving airstreams in air-conditioning systems in special situations often have to perform two functions. They may be installed in a hospital where they will be required to produce 'sterile' conditions to reduce the chances of airborne cross-infection whilst at the same time giving an acceptable thermal environment that will not stress patients who may be at risk and also allow staff to work comfortably. Some of the problems that can arise are discussed in more detail in the 'Mount Vernon Project' (p. 230).

References

1.   Holder, D. W. and North, R. S. (1963). *Schlieren Methods. Notes in applied science No. 31*. National Physical Laboratory, HMSO, London.
2.   Lewis, H. E. (1967). Colour photography of the convecting air next to the skin. *Journal of Physiology* **188,** 6–7P.
3.   Lewis, H. E., Foster, A. R., Mullan, B. J., Cox, R. N. and Clark, R. P. (1969). Aerodynamics of the human microenvironment. *Lancet* **i,** 1273–7.
4.   Clark, R. P. (1973). *The role of the human micro-environment in heat transfer and particle transport*. PhD Thesis. The City University, London.
5.   Ostrach, S. (1952). *An analysis of laminar free-convection flow and heat transfer about a flat plate parallel to the direction of the generating body force*. Technical Note 2635 (21). NACA (National Advisory Committee for Aeronautics), USA.
6.   Schlichting, H. (1960). In: *Boundary Layer Theory*. McGraw Hill, New York.
7.   Carroll, D. P. and Visser, H. (1966). Direct measurement of heat loss from a human subject. *Review of Scientific Instruments* **37,** No. 9.
8.   Nishi, Y. and Gagge, A. P. (1970). Direct evaluation of convective heat transfer coefficient by naphthalene sublimation. *Journal of Applied Physiology* **29,** 830–8.
9.   Nielsen, M. and Pedersen, L. (1952). *Acta physiologica Scandinavica* **27,** 272–94.
10.   Colin, J. and Houdas, Y. (1967). Experimental determination of the coefficient of heat exchange of the human body. *Journal of Applied Physiology* **22,** 31–38.
11.   Birkebak, R. C. (1966). *International Review of General and Experimental Zoology* **2,** 269–344.
12.   Mitchell, D. (1979). Convective heat transfer from man and other animals. In: *Heat Loss from Animals and Man*, pp. 59–76. Ed by J. L. Monteith and L. E. Mount. Butterworths, London.
13.   Clark, R. P. and Toy, N. (1975). Natural convection around the human head. *Journal of Physiology* **244,** 283–93.
14.   Froese, G. and Burton, A. C. (1957). Heat losses from the human head. *Journal of Applied Physiology* **10,** 235–41.
15.   Toy, N. (1976). *Local free and forced convection heat transfer from the human body*. PhD Thesis. The City University, London.
16.   Clark, R. P. and Toy, N. (1975). Forced convection around the human head. *Journal of Physiology* **244,** 295–302.
17.   Kerslake, D. McK. (1972). In: *The Stress of Hot Environments*. Monograph of the Physiological Society. Cambridge University Press, Cambridge.
18.   Clark, R. P., Mullan, B. J., Pugh, L. G. C. E. and Toy, N. (1974). Heat losses from the moving limbs in running: the 'pendulum' effect. *Journal of Physiology* **240,** 8–9P.
19.   Pugh, L. G. C. E. (1971). The influence of wind resistance in running and walking and the mechanical efficiency of work against horizontal or vertical forces. *Journal of Physiology* **213,** 255–76.
20.   Clark, R. P., Goff, M. R. and Mullan, B. J. (1977). Heat loss studies beneath hovering helicopters. *Journal of Physiology* **267,** 6–8P.
21.   Clark, R. P., Mullan, B. J. and Goff, M. R. (1976). *Air velocity and convective cooling coefficient measurements beneath a Sea King helicopter*. Royal Naval Personnel Research Committee (RNPRC) Report ES1/76.
22.   *Thermal Physiology and Comfort* (1981). Ed. K.Cena and J.A. Clark. Elsevier, Amsterdam and New York.

5

Heat Transfer by Radiation, Evaporation and Conduction

Radiation

Heat radiation is the term for the energy exchanged between any two bodies in the universe that have a difference in their temperature. The energy is in the form of electromagnetic radiation which can travel either through space or through gases such as the earth's atmosphere, where it may be attenuated by factors such as water vapour. Electromagnetic energy exists in a variety of wavelength bands named according to their position in the spectrum as shown in Figure 5.1. These extend from Gamma-rays, where the wavelength is of

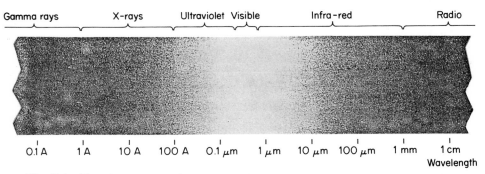

| Gamma rays | X-rays | Ultraviolet | Visible | Infra-red | Radio |

0.1 A   1 A   10 A   100 A   0.1 μm   1 μm   10 μm   100 μm   1 mm   1 cm

Wavelength

Fig. 5.1   The electromagnetic spectrum. (Reproduced with permission from AGA Infrared Systems Ltd).

the order of 0.1 ångstroms (Å)($1\text{Å} = 10^{-10}$m), to the radio waves which are extremely long at up to 100 km. The visible parts of the spectrum lie between the ultraviolet and infra-red wavelengths and the heat radiation that we are concerned with in this chapter exists in that part of the spectrum extending from the visible wavelengths (0.4–0.8 μm) to the shorter radio waves.

In discussing the mathematical properties of heat radiation it is convenient to define a black body as an object which absorbs all radiation that impinges upon it in any wavelength. The term 'black' as used here is an apparent misnomer, but is explained by Kirchhoff's Law which states that a body capable of absorbing all radiation at any wavelength is equally capable of the emission of radiation at any wavelength. The spectral distribution of radiation from a black body is dependent on the absolute temperature of the body, and

Max Planck first described this relationship by means of the following formula:

$$W_{\lambda b} = \frac{2\pi h c^2}{\lambda^5 (e^{hc/\lambda KT} - 1)} \times 10^6 \ W/m^2 \mu m.$$

$W_{\lambda b}$ = the black body spectral radiant emittance within a spectral interval 1 μm wide at wavelength λ

c   = the velocity of light = $3 \times 10^8 m/s$

h   = Planck's constant = 6.6×10^{-34} joule sec.

K   = Boltzmann's constant = 1.4×10^{-23} J/°K.

T   = the absolute temperature (°K) of the black body.

λ = wavelength (m).

The factor, 10^{-6} is used since spectral emittance is expressed in $W/m^2 \ \mu m$. If the factor is excluded, the dimension will be in W/m^2 m.

If Planck's formula is differentiated with respect to λ and a maximum is found, the following formula results:

$$\lambda_{max} = \frac{2898}{T} \mu m$$

This is Wien's formula relating the peak energy and wavelength. The spectrum of radiation may thus be characterized by either λ_{max} or T. In this formula T may be described as the colour temperature. A heated object appears red, orange and yellow as its temperature increases and the wavelength of the colour is the same as that calculated for λ_{max} at any temperature. The colour temperature of an object is important in photography where specific photographic emulsions respond to particular wavelengths, generally in the visible part of the spectrum, which are conveniently characterized using the colour temperature of the source. For example, the wavelength of radiation emitted by the sun peaks at around 0.5 μm giving a value for T of about 5,760 °K.

The quantity of heat energy emitted by a hot surface can be obtained by integrating Planck's formula from $\lambda = 0$ to $\lambda = \alpha$ to obtain the total radiant emittance (W_b) of a black body and is expressed as:

$$W_b = \sigma T^4 \ W/m^2$$

where σ = Stefan–Boltzmann constant = $5.7 \times 10^{-8} \ W/m^2 . °K^4$. This equation is known as the Stefan–Boltzmann formula and shows that the total emissive power of a black body is proportional to the fourth power of its absolute temperature. Mathematically, W_b is the area beneath the Planck curve at any particular temperature and it is of interest to note that the radiant emittance between wavelengths zero and the maximum peak is only some 25 per cent of the total. If the Stefan–Boltzmann formula is used to calculate the power radiated by the human body (at a temperature of 300 °K and having a surface area of 2 m²) a power of 1 kilowatt results. Such a large loss of energy could not be sustained and it is the compensating absorption of radiation from parts of the body which 'see' other parts and also from surrounding surfaces, generally near to room temperature, which are themselves not greatly different from the temperature of the body; the addition of clothing (having a surface temperature lower than the skin) also reduces the body's radiation heat loss.

Surfaces which are not 'black' radiate less heat than described by the Stefan–Boltzmann equation and the term 'emittance', ε, is used to define the condition of the hot surface; this is the ratio of the actual emission of heat from the surface to that radiated by a perfect 'black body' at the same temperature. A more general form of the Stefan–Boltzmann equation for a surface having any emittance is:

$$W_b = \varepsilon \sigma T^4$$

As may be seen from the above discussion any part of the electromagnetic spectrum can be involved in heat radiation. In industrial situations, where materials may be at high temperatures this is indeed true; however, in 'natural' environments there are generally two distinct bands of energy in which we should be interested. The first concerns solar radiation, which at a colour temperature of 5,760 °K, extends from the ultraviolet to the near infra-red (0.3–3.0 μm). The other band most generally of interest is from surfaces below approximately 100°C where the radiant energy is in the infra-red part of the spectrum (3–100 μm).

The wavelengths of the radiation emitted by objects such as people, plants and buildings are centred around the 10 μm band. Figure 5.2 shows the

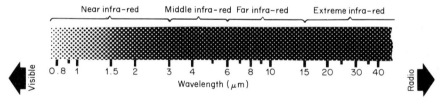

The infra-red spectrum

Fig. 5.2   The infra-red portion of the electromagnetic spectrum. Short-wave infra-red radiation occupies the 3–5 μm band with longwaves being between 8 and 14 μm. Photographic emulsion, sensitive to infra-red, generally responds to wavelengths between 0.75 and 1.2 μm in the near infra-red. (Reproduced with permission from AGA Infrared Systems Ltd.)

infra-red portion of the electromagnetic spectrum divided into the near, middle far and extreme bands depending on wavelength. The middle infra-red, between 3–5 μm, is the wavelength band exploited by infra-red thermography described in Chapter 2.

The next problem to be encountered in discussing the radiation exchanges between surfaces and objects in enclosures (for instance a man standing in a room) is to establish the mathematical relationship between the net radiant energy emitted and absorbed by each surface. Kerslake[1] reviews the radiation exchange between surfaces and shows that the radiant energy transfer between two parallel surfaces (A and B) of infinite size can be described by the formula:

$$R = \frac{\sigma(T_A^4 - T_B^4)}{(1/\varepsilon_A + 1\varepsilon_B - 1)}$$

In the case of a body within an enclosure the mathematical analysis is somewhat complex and can only be rigorous for relatively simple shapes. For concentric cylinders and spheres the following equation by Christiansen applies. This may be used as an approximation for the radiant heat transfer from any object (1) to any enclosure (2) having areas A_1 and A_2 respectively.

$$R = \frac{\varepsilon_1 \sigma(T_1^4 - T_2^4)}{1 + \varepsilon_1(1/\varepsilon_2 - 1)} \frac{A_1}{A_2}$$

If Christiansen's equation is applied to a very large enclosure $(A_2 \gg A_1)$ the equation approaches:

$$R = \sigma\varepsilon_1(T_1^4 - T_2^4)$$

In this case the emittance of the enclosure does not influence the radiant heat exchange at all, as energy from the object reflected back by the enclosure generally misses the object and strikes other parts of the enclosure. If an emittance factor F_e is introduced the general equation for radiative heat transfer becomes:

$$R = F_e \sigma(T_1^4 - T_2^4)$$

Kerslake has pointed out that the term $(T_1^4 - T_2^4)$ is rather laborious to evaluate and he has worked out values of σT^4 at various temperatures in °C. Table 5.1 shows these values. For example the radiation exchange between

Table 5.1   Black body radiation at various temperatures. (Reproduced with permission from Kerslake (1972)).[1]

°C	Values of σ. T^4(W/m²) at different temperatures (°C)									
	0	1	2	3	4	5	6	7	8	9
0	315	320	325	329	334	339	344	349	353	358
10	364	369	374	379	385	390	396	401	407	413
20	418	424	430	436	442	448	453	459	465	472
30	478	484	491	497	504	511	517	524	531	537
40	544	551	558	565	573	580	587	595	602	610
50	618	625	633	641	649	656	664	673	681	689
60	697	706	715	723	732	741	750	758	767	776
70	785	794	804	813	823	832	842	851	861	871
80	881	891	901	911	922	932	943	952	963	974
90	985	996	1007	1018	1028	1041	1052	1063	1075	1086
100	1098	1110	1122	1134	1147	1158	1170	1183	1195	1208

surfaces at 50°C and 30°C, $\sigma(T_1^4 - T_2^4) = 618 - 478 = 140 \text{ W/m}^2$. The radiant exchange in any configuration will be found by multiplying this value by the appropriate emittance factor which is the quantity depending on the emittances of the two surfaces involved.

In the case of heat transfer from man, not all of the body area takes part equally in the radiative process and the equation may be written as:

$$R = \sigma(T_1^4 - T_2^4)A_r/A$$

where the emissivity for human skin is approximately equal to 1 and where A is the total body surface area and A_r the area exchanging radiation energy with the surroundings[2]. The ratio A_r/A is dependent on posture and has a value for a naked man in a spread-eagled position of 0.95; during cycling the value is 0.81 and 0.78 when reclining. Fanger[3] lists a coefficient f_{cl} (the ratio of the surface area of the clothed to nude body) which enables A_r/A to be evaluated for various clothing assemblies.

The radiation area A_r can be determined in several ways; with partitional calorimetry either the total sensible heat loss at various air movements can be measured and then extrapolated to zero convective heat loss, or the radiant temperature of an enclosure can be varied independently of air temperature and the effect of the radiant heat exchange compared to that calculated for the whole DuBois area.

Another approach is to photograph the subject from different directions to obtain values of the projected area and to integrate these over all possible directions[4]. Underwood and Ward[5] produced silhouettes of subjects (Fig. 5.3) photographed from various angles of altitude and azimuth from which to calculate this area.

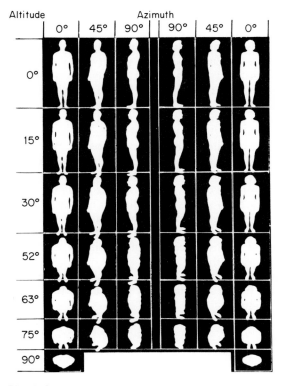

Fig. 5.3 Direct radiation area of male (three columns on left) and female (three columns on right) subjects at altitudes of 0°–90°, at three angles of azimuth. (Reproduced with permission from Underwood and Ward (1966)[5]).

Radiation calculations simplified

Evaluating the equation for radiant heat exchange is rather involved because the temperature term is raised to the fourth power. It is sometimes more convenient to express radiation exchange by a first power relationship together with a radiation coefficient. The expression $R = F_e \sigma (T_1{}^4 - T_2{}^4)$ would then be equivalent to $F_e\, h_r(T_1 - T_2)$ with h_r having the units of $W/m^2.°C$. For the case of a naked subject with a mean skin temperature of 33°C in a range of room temperatures the value of h_r lies between 6 and 7 $W/m^2.°C$. Kerslake discusses this in detail and arrives at the value of 6.70 $W/m^2.°C$ for h_r which is appropriate to most environments. The calculation of body radiation heat exchange then becomes similar in form to that for convection and evaporation. When the radiation area expressed as a ratio of the DuBois surface area is taken into account ($A_r = 0.81\mathring{A}$ on average for a standing man) in the first power relationship the resulting coefficient for radiation exchange becomes $h_r = 5.4$ $W/m^2.°C$ where the area is calculated according to the DuBois formula.

Direct solar radiation

In bright sunlight, solar radiation on the body may be the largest factor in the heat balance equation and can be as much as ten times the metabolic heat production. This is short wave radiation ($0.3-3.0\ \mu m$) which is incident upon the subject and where subsequently a fraction will be lost by reflection. The formula for estimating the solar radiant heat gain is $H_r - A_p \alpha_s I_s$ where A_p is the radiation area for the prevailing solar angle, α_s is the absorptance of the skin or clothing in sunlight (the proportion of the incident solar radiation absorbed) and I_s is the radiation flux (W/m^2) normal to the radiation. The direct radiation area for an erect standing body was evaluated by Ward and Underwood[6] as $A_d = 0.0429 \sin \theta + 0.385 \cos \theta$ where θ is the solar altitude. The constants apply for an average body area of 1.8 m², and for subjects of different area A, they must be multiplied by A/1.8. Similar areas for other postures are also available. Figure 5.4 shows the radiation area as the percentage of the DuBois area for male and female subjects at different solar altitudes. When subjects are thickly clothed, Pugh and Chrenko[7] have shown that the ratio of projected area to total surface area of clothing is practically equal to the equivalent ratio in nude subjects.

Estimates of the solar heat load normal to the sun's rays can be obtained by multiplying the shadow area by the solar intensity measured with the flat surfaced radiometer held horizontally.

Indirect solar radiation

Athough by far the greatest radiant heat load comes from direct solar radiation, there is an additional component due to reflections both from the ground or clouds and from sunlight that is scattered throughout a clear sky. Measurements of the indirect radiation indicate an increase of between 20 and 30 per cent in the net solar radiation flux over the direct radiation[8, 9].

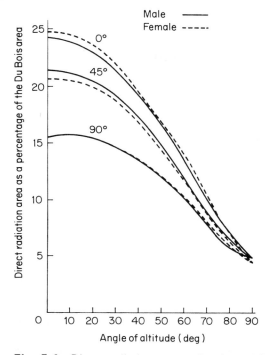

Fig. 5.4 Direct radiation areas of male and female subjects at three angles of azimuth (0° corresponds to facing the sun) and at solar altitudes of 0°–90°. The radiation areas of males and females expressed as a percentage of the DuBois surface area are similar. (Reproduced with permission from Underwood and Ward (1966)[5].)

The greenhouse effect

Air temperatures in a greenhouse are in general higher than those of the ambient outside air even if there is only a small amount of sunlight. This heating effect is due to the trapping of energy within the greenhouse. Ordinary window glass is highly transparent to short wave high temperature solar radiation. When the light strikes the glass it passes through with hardly any attenuation and the objects within the greenhouse absorb the radiant energy from this light. Everything inside, the plants, the tools, the benches, etc. become warmed by this absorption of radiant energy and consequently their surface temperatures increase. The ambient air within the greenhouse then becomes heated by convection from these objects as illustrated in Figure 5.5. The warm surfaces will also re-radiate energy but in the relatively long wave part of the electromagnetic spectrum, that is beyond the infra-red. The glass is practically opaque to this long wave radiation and there is thus an energy trap in which the incoming radiant energy is, in effect, stored within the greenhouse to produce higher temperatures. The surfaces which re-radiate do so, not only to the glass itself but also to all of the other objects within the

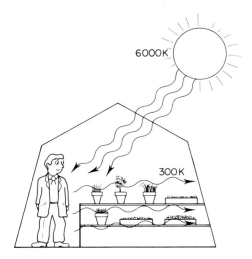

Fig. 5.5 Diagram to illustrate the 'greenhouse' effect by which glass acts as a one-way trap for the sun's radiant energy. The wavelengths of radiation from the sun are mostly short and pass through the glass to heat objects within the greenhouse. The longer wavelength radiation from the heated objects within the greenhouse are blocked by the glass from leaving.

greenhouse. There is radiation bouncing back and forth which leads to a further exchange of energy between these objects and the general raising of their temperature, which in turn produces an increased convective heat exchange to the air within the greenhouse. This effect will also apply to buildings with ordinary windows and also to the interiors of motor cars even if they have a relatively small window area. As far as buildings are concerned, this effect can make some of them, with large window areas, extremely uncomfortable.

The emission curves at solar and terrestrial temperatures compared with the spectral transmission curve for ordinary window glass are seen in Figure 5.6. Note that the air inside an empty glass sphere will not heat up, and there will be no 'greenhouse effect' since the short wave solar radiation will pass through without being trapped because there is no absorbing material inside.

Solar radiation and skin temperatures

An examination of the temperature changes that occur during sunbathing, using infra-red thermography, gives some insight into the physiological and physical processes by which the body responds to solar radiation.

Thermographic visualization has been made of skin temperatures before and during 'sunbathing' with the subject standing and facing directly into the sun (dry and wet bulb shade temperatures of 31 and 19°C respectively with no unevaporated sweat visible on the skin). A feature of the resting shade temperature distribution before sunbathing was the absence of the patterns reflecting the structures beneath the skin. Large areas of the skin surface were

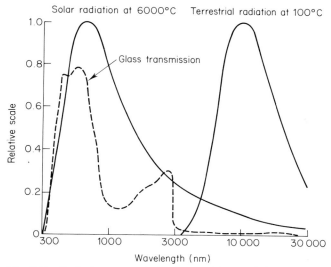

Fig 5.6 Emission curves at solar and terrestrial temperatures compared with a spectral transmission curve for ordinary window glass. (Reproduced with permission of J. Longmore, School of Environmental Studies, London.)

at a uniform temperature and this was similar to the pattern described in the section dealing with high environmental temperatures (Chapter 3). After the subject had been standing in the sun for some 20 minutes the temperatures reached a steady state. There was a marked difference between frontal areas facing directly into the sun and those over the back which were in shade; skin facing the sun had risen by some 5°C overall whereas the back temperature had fallen by about 1°C. The temperatures of 'passive' surfaces, such as the bathing trunks and the hair, had increased substantially to 45–48°C.

This redistribution of temperature illustrates the ability of the body to thermoregulate by varying the thermal properties of the skin. Changes in blood supply and sweat rate on surfaces facing into the sun can increase net heat loss to such an extent that the solar load only increases these temperatures by some 5°C. Increased sweating over the shaded areas of the back increased heat loss, and without direct solar load the temperatures here were able to fall due to the increased evaporation. The swimming trunks and the hair did not of course have the ability to vary heat gain or loss and simply increased in temperature on exposure to direct sunlight. Previous observations of the influence of direct solar load[10] have shown skin temperatures as high as 40°C. These observations were made at lower ambient temperatures than in the thermographic investigation, and a lower heat loss by evaporation, because of less sweating, could account for the higher skin temperature observed.

In the thermographic observations a definite division between the temperature pattern on the front surface of the face and the rest of the body was found. The face temperatures were about 4°C below those over the rest of the trunk although the whole of the front of the body was facing the sun. This demarcation of face and trunk has been seen in other environments (Chapter 3) but in this case could possibly be accounted for by a greater sweat rate than over the rest of the trunk.

UV-A, B and C radiation

The biological effects of radiation between short ultraviolet and visible wavelengths are currently receiving considerable attention. This has stemmed from concern about safe exposure limits in industrial processes as well as from the increasing use of ultraviolet radiation for the treatment of clinical conditions and for cosmetic and 'sun-tanning' purposes.

The term 'photobiology' is now widely used for the study of the effects of ultraviolet (UV) and visible radiation on biological material; the effects of ionizing radiation (shorter than 200 nm) fall into the province of radiobiology.

The term 'electromagnetic' is used for the wide spectrum of radiation produced by the sun because it is propagated in the form of oscillatory electric and magnetic fields. As well as having a wave-like nature, however, this radiation also has properties similar to those of discrete particles or packets of energy which are termed 'quanta' or 'photons'. Planck showed that the energy of a photon was directly proportional to its frequency and consequently shorter wavelength; higher frequency, electromagnetic radiation consists of high energy photons.

The UV portion of the spectrum is fairly arbitrarily divided into three parts designated UV-A, UV-B and UV-C. Wavelengths between 200 and 290 nm are in the UV-C range; radiation shorter than 200 nm is mainly absorbed by the air and that below 290 nm by ozone in the upper atmosphere. The UV-B radiation occupies the band between 290 and 320 nm and that between 320 and 400 nm is the UV-A region.

Not a great deal of UV-B reaches the earth's surface but nevertheless it is very efficient at causing sunburn and has been shown to produce cancer in laboratory animals and to change the molecular structure of bacteria to produce mutations. There is therefore a strong suggestion that these wavelengths can cause skin cancer and ageing in man.

Hitherto UV-A radiation has been considered harmless; it does produce pigmentation or tanning as well as redness of the skin although the energy required for this is some orders of magnitude greater than for radiation in UV-B region. The maximum irradiance in the UV-A wavelengths is about 50 W/m^2 at sea level and may be greater at higher altitudes.

Besides the demands of holiday-makers for methods to safely extend the time of exposure to the sun, there are numerous 'treatments' offered to prepare for the holiday and to maintain the 'tan' long into the winter months. This has resulted in the proliferation of sun-beds and solaria which can be bought or hired, often with a minimum of supervision or instruction. These machines frequently have the claim of being completely safe because all of the UV-B rays are filtered leaving only those in the UV-A region. Such claims are now under scrutiny and there is a real need for standards, and possibly regulations, in the use of artificial sources of UV-A. High doses may potentiate the more biologically active UV-B, particularly in regard to the effect on cells and micro-organisms. Sunscreens do not significantly attenuate the amount of UV-A reaching the skin and window glass and many plastics pass this radiation whilst not transmitting UV-B. In addition, UV-A is now widely used in industrial photochemical processes and is being investigated in the treatment of chronic skin disorders (such as psoriasis and eczema) in conjunction with photosensitizing drugs.

Radiation and the structure of the skin

It is beyond the scope of this book to give more than an outline of the complex relationships between the structure of the dermis and epidermis and the way that incident radiant energy is handled by the skin. Special cells known as melanocytes are found within the basal layer of the epidermis; these cells produce pigment granules containing melanin which is a complex protein that strongly absorbs light and UV radiation and which ultimately becomes deposited in the stratum corneum. Skin colour differences between races are due to different rates of melanin production and not necessarily to the number of melanocytes.

Some protection against sunburn is provided by the UV absorbing properties of the melanin although the exact mechanisms for this are not entirely clear and it is now suggested that melanin granules are not simply acting as optical filters. Studies of electron spin resonance have shown that black and white skin, when irradiated with UV (250–320 nm) produce quite different signals. The greater response appears to be induced by the UV stimulus in white skin[11].

Much work needs to be done on this topic but it is possible that melanin may act as an electron 'trap' as well as being an optical absorber.

Cutaneous blood flow

In addition to transporting oxygen, nutrients and defensive material and removing products of metabolism, the cutaneous vascular system also performs a vital role in body temperature regulation.

Subcutaneous arteries have branches which form a hypodermal arterial plexus which supplies blood to the hair follicles and sweat glands. Other branches form dermal and sub-epidermal plexuses which have short interconnections. In some areas, notably the hands, feet, lips, nose and ears, there is communication between the arteries and venous plexuses known as shunts or arteriovenous anastomoses. The walls of these vessels have muscle coats together with sympathetic vaso-constrictor nerve fibres. When dilated, they release a large blood flow to the plexuses and this is reduced to practically zero when the walls are constricted. These anastomoses can therefore produce a rapid response to thermal stress on the skin surface.

Skin colour

Human skin varies in its reflectance properties when subjected to short wave radiation. The Negroid skin reflects some 16 per cent of the energy in these wavelengths whilst white skin reflects some 30–40 per cent[13]. Black skin reflects less visible or ultraviolet wavelengths but sun-tanning of white skin reduces reflectance of the short wavelengths. Over the whole spectrum, white skin absorbs some 60 per cent of the incident radiation compared with 80 per cent in the Negro. However, dark skin allows less penetration of solar radiation than white skin which compensates for the reflection disadvantage. Negroid skin contains more pigment which acts as a filter and limits the penetration to a fraction of a millimetre compared with the several millimetres penetration in white skin. Figure 5.7 shows the effect of solar flux on the absorptance of white and Negroid skin.

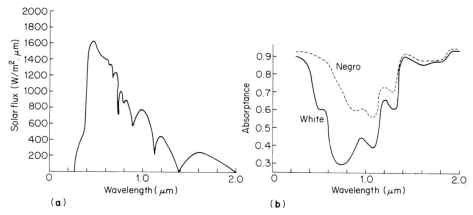

Fig. 5.7   Spectral distribution of solar radiation (a) and of absorptance of white and Negroid skin (b). (Reproduced with permission from Kerslake (1972)[1], redrawn from Gates (1962)[14] and Jacquez *et al*. (1955)[15, 16].)

It may well be asked what advantage, if any, there is in a dark or a black skin. It would be expected that skin colour would have some biological significance, as it is in hot countries with a high incidence of solar radiation that the indigenous people generally are dark or black. It is at present surprisingly difficult to find any obvious physiological advantage in having such a pigmented skin, except that some of the radiation is absorbed in the pigmented melanin layer and the subject is therefore less likely to develop sunburn; otherwise there is no evident advantage. It may be added that people with pigmented skins very rarely develop rodent ulcers on the skin, whereas non-pigmented people living in hot climates with high solar radiation tend to have a very high incidence of these ulcers. Nevertheless, most people who have considered this still have a degree of puzzlement over the advantage or disadvantage of a black skin. It is felt that something here has been missed but quite what this is remains uncertain.

Skin emissivity

A number of workers[17, 18] have studied the emissive properties of human skin in the infra-red region both in vivo and in vitro. For normal intact skin the average emissivity was found to be between 0.97 and 0.98 irrespective of skin colour or pigmentation and is therefore very like a perfect 'black body'. Most clothing materials also have emissivities near to unity. Table 5.2 (pages 110 and 111) is useful for comparing the emissivities of various materials particularly when evaluating radiation exchanges between the human body and a surrounding enclosure.

Evaporative heat loss

Evaporation is an extremely important route of heat loss and it is the main mechanism by which the human body avoids overheating in warm conditions. Dry heat exchange, by radiation and convection, decreases as the ambient temperature rises thereby producing a smaller temperature difference between the skin and the surroundings. In order to avoid energy storage and ultimate overheating in such conditions excess heat has to be dissipated by the evaporation of sweat. Similarly, during strenuous activity, such as running, evaporative cooling plays a dominant role in thermal balance by maintaining body temperature within acceptable limits. Evaporative cooling and sweat rate can be finely adjusted over a wide range of energy production. For instance, metabolic heat production can vary by more than one order of magnitude from about 100 W at rest to over 1 kW during violent exercise. In hot conditions, sweat rates can be as high as 1.5 litres/h in unacclimatized persons and up to 4 litres/h for someone fully acclimatized to heat. The corresponding maximum heat energy that could be lost at these sweat rates are 1 kW and 2.7 kW respectively. For each gram of water that evaporates from the body surface 0.58 Kcal of heat are extracted. This cooling effect is lost if the sweat runs off as liquid without evaporating.

In the absence of active sweating the body can still lose heat by evaporation from water lost by diffusion through the skin; from the respiratory tract water loss occurs at the rate of approximately 0.5 g/min to give a heat loss of about 20 W.

The heat loss by evaporation will depend on the rate at which the body can secrete sweat and on the evaporative capacity of the environment. If environmental conditions are near to saturation it may not be possible for the air to absorb water vapour at the rate at which it is produced at the skin surface and consequently the heat extraction from the body will be lower. On the other hand, if the sweat secretion rate is less than maximal, or if there are regional variations over the body surface, evaporative cooling will not be at a maximum.

The processes of convection and evaporation have much in common. Convection is a diffusion of heat across the natural convection boundary layer surrounding the skin whilst evaporation is the diffusion of water vapour across the boundary layer once the liquid sweat has been converted to the vapour phase.

In convection, the heat transfer is determined by the temperature gradient across the boundary layer and in evaporation, the vapour concentration gradient determines the mass transfer of vapour ($\dot{m}$). This may be described by the equation:

$$\dot{m} = h_d \, (C_s - C_a) \; g/m^2 \, s.$$

Table 5.2 Emissivities (total normal) of various common materials.

Materials	Temperature (°C)	Emissivity (ε)
Metals and their Oxides		
Aluminium:		
polished sheet	100	0.05
anodized sheet, chromic acid process	100	0.55
Brass:		
highly polished	100	0.03
oxidized	100	0.61
Copper:		
polished	100	0.05
heavily oxidized	20	0.78
Gold: highly polished	100	0.02
Iron:		
cast, polished	40	0.21
cast, oxidized	100	0.64
sheet, heavily rusted	20	0.69
Silver: polished	100	0.03
Stainless Steel (types 18–8):		
Buffed	20	0.16
oxidized	50	0.85
Steel		
polished	100	0.07
oxidized at 800°C	200	0.79
Tin: commercial tin-plated sheet iron	100	0.07
Other materials		
Brick: common red	20	0.93
Carbon:		
candle soot	20	0.95
graphite, filed surface	20	0.98
Concrete:	20	0.92
Glass: polished plate	20	0.94
Oil, lubricating (thin film on nickel base):		
nickel base alone	20	0.05
film thickness 0.025 mm	20	0.27
0.051 mm	20	0.46
0.125 mm	20	0.72
thick coating	20	0.82
Paint, oil: average of 16 colours	100	0.94
Paper: white bond	20	0.93
Plaster: rough coat	20	0.91
Sand	20	0.90

Materials	Temperature (°C)	Emissivity (ε)
Skin, human	32	0.98
Soil:		
dry	20	0.92
saturated with water	20	0.95
Water:		
distilled	20	0.96
ice, smooth	−10	0.96
frost crystals	−10	0.98
snow	−10	0.85
Wood: planed oak	20	0.90

where C_s and C_a are the mass/unit volume water vapour concentrations at the skin surface and ambient air respectively and h_d is known as the mass transfer coefficient with the units of ms^{-1}.

In considering the evaporative heat loss from the body it is more convenient to relate the evaporation rate to the equivalent heat transfer (E) and to use the water vapour pressure (p_s and p_a for the skin and air respectively) instead of vapour concentration. This yields the equation:

$$E = h_e (p_s - p_a)$$

where h_e is the evaporative coefficient with units of $W/m^2 .°C \, kPa$.

Direct determinations of the evaporative coefficient (h_e) rely on measuring the evaporation rates of subjects with completely wet skin in environments with fairly high water vapour pressure which makes the measurement difficult. Kerslake[1] has summarized the results of various workers who have determined h_e at different postures and air movements (V − m/s) and these are shown in Table 5.3. He concludes that for all practical purposes the formula:

$$h_e = 124 \sqrt{V} (W/m^2 . kPa)$$

may be used and that the ratio $h_e/h_c = 15$ is consistent with observations on human subjects.

Table 5.3 Evaporative coefficients according to various authors.

Author	Range of airspeed V-m/s	Formula W/m².kPa	Position
Buettner, 1934[19]	0.15–0.50	$h_e = 120 \, V^{0.5}$	Lying on bed
Nelson *et al.*, 1947[20]	0.15–3.00	$h_e = 88 \, V^{0.37}$	Standing
Clifford *et al.*, 1959[21]	0.58–4.00	$h_e = 109 \, V^{0.63}$	Standing

Heat transfer by conduction

In solids, heat flows as a result of the transfer of thermal energy from molecule to molecule. This process is called conduction and also occurs in liquids and gases where the molecules are not constrained but can move around. In this way the conduction process is the first mechanism involved in the initiation of convective heat transfer (described in Chapter 4). Figure 5.8 represents a plane wall with temperatures T_1 and T_2 and thickness b.

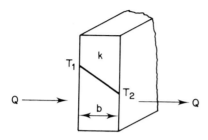

Fig. 5.8 Steady state heat conduction Q, through a wall of thickness b, thermal conductivity k, with wall temperatures of T_1 and $T_2 = k(T_1 - T_2)/b$ per unit area.

The quantity of heat that flows through an area A of the wall in unit time is Q and is given by Fourier's law as $Q = kA(T_1 - T_2)/b$ where k is the thermal conductivity of the wall material (W/m.°C). The temperature within the wall decreases linearly from T_1 to T_2 when conductivity is not temperature dependent. Table 5.4 shows the thermal conductivity of several materials.

Table 5.4 Thermal conductivity of various materials.

Material	Thermal conductivity (W/m.°C)
Copper	400
Aluminium	210
Glass	1.1
Concrete	0.92
Brick	0.63
Water	0.59
Cork	0.046
Air	0.025
Thick fabrics	0.04
Body tissue, muscle	0.39
Body tissue, fat	0.20

Architects and engineers sometimes find it convenient for calculating heat loss to define a 'U' value for the structure as a whole which takes account of the conductivity and the normal thickness of the structure. The 'U' value is defined as the rate of heat flow per unit area, per degree of difference of temperature between surfaces. Some examples are given in Table 5.5.

Table 5.5 The 'U' value of different structures.

Structure	'U' value (W/m².°C)
Solid double brick wall	2.8
Similar wall but plastered one side	2.5
Plate glass (6 mm)	10.0
Plastered ceiling	2.0

In a composite solid, consisting for example of three layers, as in Figure 5.9 with different conductivities k_1, k_2 and k_3, the equations for heat transfer through the various layers are:

$$T_1 - T_2 = b_1 \, Q/k_1 \, A$$
$$T_2 - T_3 = b_2 \, Q/k_2 \, A$$
$$T_3 - T_4 = b_3 \, Q/k_3 \, A$$

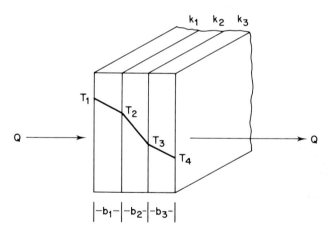

Fig. 5.9 Steady state heat conduction through a composite plain wall with various temperatures (T) and thickness (b) and thermal conductivity (k) of materials.

Addition of these equations gives:

$$T_1 - T_4 = (b_1/k_1 A + b_2/k_2 A + b_3/k_3 A) \, Q$$

from which Q may be calculated from the surface temperatures T_1 and T_4.

The quantity of heat lost by the adult human to solid surfaces such as chairs, beds and floors by direct conduction is generally low because the areas of contact are small and the thermal conductivity of the support material is poor. Wood and upholstery have low thermal conductivities and generally do not give rise to discomfort. Materials such as metals, solid plastics and stone with higher conductivities may produce local discomfort when they are cold. Babies nursed in incubators have some 10 per cent of their total surface area in contact with the support tray and direct conduction can account for up to 20 per cent of the total heat loss. This can be reduced to around 3 per cent if the infant lies on a 1 cm thick foam mattress[22].

Conduction from the body 'core'

In calculations of body heat balance a useful concept is that of a central body 'core' where the so-called vital organs of the body are maintained at a constant temperature.

Tissues surrounding this 'core' can be said to form a shell consisting of skin, underlying fat and superficial layers of muscle. Depending on the metabolic activity in the 'core' and the heat exchange processes at the skin surface in relation to the environment, the shell will vary in temperature and thickness. Fox[23] represented the 'core' as a shaded area, different for hot, neutral and cold environments as shown in Figure 5.10. A simplified way of dealing with calculations of body heat exchange with the environment is to consider the 'core' covered by tissues and clothing as analogous to a hot vessel surrounded by lagging.

Heat is conducted from the core through the tissues of the shell to give a mean surface temperature in 'comfortable' conditions of around 33–34°C. When the body is covered with clothing there is further insulation and the outer surfaces of the garments are intermediate between skin and environ-

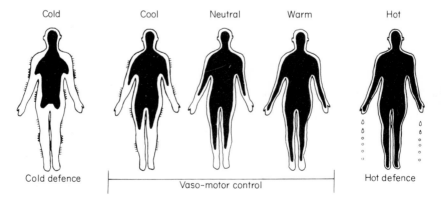

Fig. 5.10 Representation of the size of the central constant temperature 'core' in conditions ranging from hot to cold. In cool, neutral and warm conditions temperature regulation is affected by vaso-motor control. In hot conditions, sweating becomes necessary and in the cold, shivering is evoked.[23]

mental temperature depending on the insulation properties of the fabric. In steady equilibrium conditions, body heat exchange with the surroundings can be considered in two distinct parts. First there is the heat conducted through the tissues and clothing, and second this energy is lost from the surface to the environment. These separate heat exchanges can be represented graphically as in Figure 5.11. Heat conduction through the shell, tissues and clothing falls

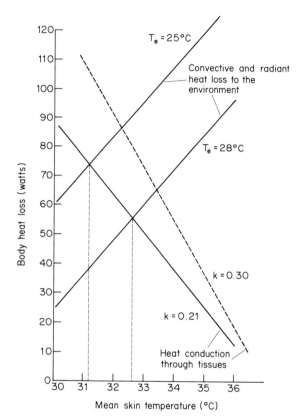

Fig. 5.11 Heat loss through body tissues by conduction from the 'core' plotted against skin surface temperatures for two values of tissue conductance k. When these lines meet the curves of heat loss from the surface to the environment, the conduction through the tissues is equal to the heat transfer by radiation and convection to the surroundings. In this state the equilibrium surface temperature can be read from the intersection of the lines. In this example the body is treated as a 'lagged' pipe and the core temperature has been taken as 37°C, the tissue thickness surrounding the core as 2.4 cm (Crosbie, Hardy and Fessenden (1963)[24]) and the tissue conductance given average values of 0.21 W/m.°C (solid line) and 0.30 W/m.°C for fat and muscle (dashed line). Lines for environmental heat loss are shown for 25°C and 28°C where the resultant mean surface temperatures are 31.2°C and 32.6°C respectively shown by the vertical dotted lines intercepting the temperature axis.

if the clothing temperature increases; in this situation the heat exchange with the environment by radiation and convection will rise. A steady state will exist when the heat conducted through the tissues and clothing is equal to that lost to the environment. The resulting surface temperature will be found where the two heat exchange lines cross on the diagram. This analogy with a lagged pipe is useful in understanding why insulation that can be provided for the fingers, for example, is limited, and why they are often cold when wearing gloves even though the insulation value of the material may be high. When small diameter cylinders are lagged the surface area exchanging heat with the environment is substantially increased and this can outweigh the advantage of greater thermal insulation with its attendant decrease in surface temperature. In such a case it is possible for the insulated fingers to actually lose more heat when covered by gloves than when bare.

When single cylinder manikins are used for body heat transfer calculations their diameter should be at least 18 cm. With smaller models, errors in measurement and calculation can arise particularly when assessing the insulation of fabrics because of this 'curvature' effect.

In small animals, such as mice or rats, the same reasoning can show that the fat animal with a thicker insulation around a central core can lose more heat to the surroundings than a lean one. Such a phenomenon can be misinterpreted as increased metabolism in the fat animal and although such effects are familiar to physicists they have not always been appreciated by biologists and physiologists.

References

1.   Kerslake, D. McK. (1972). *The Stress of Hot Environments*. Monograph of the Physiological Society. Cambridge University Press, Cambridge.
2.   Mitchell, D., Wyndham, C.H., Vermeulen, A.J., Hodgson, T., Atkins, A.R. and Hofmeyr, H.S. (1969). Radiant and convective heat transfer of nude men in dry air. *Journal of Applied Physiology* **26**, 111–18.
3.   Fanger, P.O. (1967). Calculation of thermal comfort: introduction of a basic comfort equation. *Transactions of the American Society of Heating, Refrigeration and Air-conditioning Engineers* **73**, 1–16.
4.   Guibert, A. and Taylor, C.L. (1952). Radiation area of the human body. *Journal of Applied Physiology* **5**, 24–57.
5.   Underwood, C.R. and Ward, E.J. (1966). The solar radiation area of man. *Ergonomics* **9**, 155–68.
6.   Ward, E.J. and Underwood, C.R. (1967). The effect of posture on the solar radiation area of man. *Ergonomics* **10**, 399–410.
7.   Pugh, L.G.C.E. and Chrenko, F.A. (1966). The effective area of the human body with respect to direct solar radiation. *Ergonomics* **9**, 63–7.
8.   Roller, W.L. and Goldman, R.F. (1968). Prediction of heat load on man. *Journal of Applied Physiology* **24**, 717–21.
9.   Lee, D.H.K. (1964). Terrestrial animals in dry heat: Man in the Desert. In: *Handbook of Physiology*, Section 4, pp. 551–82. Ed by D.B. Dill. American Physiological Society, Washington.
10.   Pugh, L.G.C.E. (1964). Solar heat gain by man in the high Himalayas. In: *Environmental physiology and psychology in arid conditions*. Unesco Symposium, 7–12 Dec. 1962.

11.   Pathak, M.A. and Stratton, K. (1968). Free radicals in human skin before and after exposure to light. *Archives of Biochemistry and Biophysics* **123**, 458.

12.   Marples, M.J. (1965). *The Ecology of the Human Skin*. Charles C Thomas, Springfield, USA.

13.   Tregear, R.T. (1966). In: *Physical functions of skin*. Academic Press, London and New York.

14.   Gates, D.M. (1962). *Energy Exchange in the Biosphere*. Harper and Ross, New York.

15.   Jacquez, J.A., Huss, J., McKeehan, W., Dimitroff, J.M. and Kupperheim, H.F. (1955). Spectral reflectance of human skin in the region 0.7–2.6 μm. *Journal of Applied Physiology* **8**, 297–9.

16.   Jacquez, J.A., Kupperheim, H.F., Dimitroff, J.M., McKeehan, W. and Huss, J. (1955). Spectral reflectance of human skin in the region 235–700 μm. *Journal of Applied Physiology* **8**, 212–19.

17.   Mitchell, D., Wyndham, C.H. and Hodgson, T. (1967). Emissivity and transmittance of excised human skin in its thermal emission wave band. *Journal of Applied Physiology* **23**, 390–4.

18.   Patil, K.D. and Lloyd-Williams, K.L. (1969). Spectral study of Human Radiation. In: *non-ionising radiations* **1**, 56. Iliffe Science and Technology Publications Ltd, London.

19.   Buettner, K. (1934). Die Wärmeübertragung durch Leitung und Konvektion. *Verdunstung und Strahlung in Bioklimatologie und Meteorologie*. Veröffentlichungen des Preussischen Meteorologischen Instituts. Abhandlungen Bd. x No. 5, Berlin.

20.   Nelson, N., Eichna, L.W., Horvath, S.M., Shelley, W.B. and Hatch, T.F. (1947). Thermal exchanges of man at high temperatures. *American Journal of Physiology* **151**, 626–52.

21.   Clifford, J.C., Kerslake, D. McK. and Waddell, J.L. (1959). The effect of wind speed on maximum evaporative capacity in Man. *Journal of Physiology* **147**, 253–9.

22.   Hey, E.N., Katz, G. and O'Connell, B. (1970). The total thermal insulation of the new born baby. *Journal of Physiology* **207**, 683–98.

23.   Fox, R.H. (1974). Temperature regulation. In: *Recent Advances in Physiology*, Chapter 8. Ed by R.J. Linden. Churchill Livingstone, Edinburgh and London.

24.   Crosbie, R.J., Hardy J.D. and Fessenden E. (1963). Electrical analog simulation of temperature regulation in man. In: *Temperature: Its Measurement and Control in Science and Industry*. Ed. J.D. Hardy. Part III, pp 627–35. Rheingold, New York.

6

Temperature and the Circulation

The term 'circulation' is used to describe the pumping action of the heart together with the system of vessels through which the blood flows. Body temperature depends, to a considerable extent, on the circulatory system and its various control mechanisms; environmental temperature also affects the action of the heart and blood vessels.

The main function of the circulatory system is to enable blood to carry oxygen to the tissues and to remove CO_2 and other metabolites from them. The first and most obvious temperature effect on the circulation is a physical one relating to the viscosity of the blood; the lower the temperature, the greater the viscosity. This leads to an important haemodynamic consequence for blood pressure, which depends upon the product of cardiac output and peripheral resistance, changed mainly by the diameter of the arteries and arterioles. If these vessels are constricted the resistance to flow and the blood pressure increase and if they are dilated the resistance and pressure fall. Changes in viscosity affect the peripheral resistance and if blood temperature falls then viscosity increases as will the resistance to flow. This can be very marked in cold conditions when the greater local resistance reduces the rate of blood flow. Temperature is also important in determining the relationship between respiration and blood flow. The dissociation curve (Fig. 6.1) relating

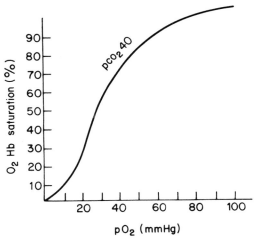

Fig. 6.1 Oxygen Hb dissociation curve at a CO_2 tension of 40 mmHg.

oxygen tension and the quantity of oxygen carried in the blood in combination with haemoglobin, is dependent on temperature; the curve is shifted to the left in the cold and oxyhaemoglobin dissociates less freely as oxygen pressure falls. At low temperatures, the blood does not give up oxygen as readily at low partial pressure of oxygen as at normal blood temperature; and at very low temperatures the dissociation is markedly affected and oxygen is only released from combination with haemoglobin at very low partial pressures. These effects are mainly evident in the periphery and specifically in the skin of the extremities such as the hands, feet and to some extent the face, notably the nose and ears. Temperature is also important for all other physical and chemical changes in the blood.

Cardiac effects

Changes in temperature influence heart rate, and therefore may alter cardiac output. If body temperature is raised, heart rate increases by approximately ten beats per minute for each 1°C rise. There is an increased cutaneous blood flow, but relatively little change in cardiac output. In vitro studies on isolated heart lung preparations have demonstrated that the output per beat increases as the temperature of the perfusate is raised. This effect is not observed in intact man. When body temperature is lowered, cardiac output per minute is reduced since the heart rate is lower. Blood flow during each beat does not appear to change dramatically until body temperature has fallen to about 33°C. At this temperature and below there are increasing changes in cardiac output which diminishes, not only because of the fall in heart rate, but because of the reduced output per beat which is largely due to increasing blood viscosity. The viscosity effects become rather marked as the whole blood volume is involved, not simply the blood contained in the superficial vessels which are exposed to direct cooling. At very low body temperatures, that is below 32°C, there is an increasing likelihood that the heart beat will become irregular with extra systoles. When the temperature falls to around 28°C and below, there is an increasing risk of ventricular fibrillation, which causes a calamitous fall of blood pressure with death resulting within a matter of seconds. Ventricular fibrillation is the real danger in severe hypothermia, and when such cooling occurs in the course of a surgical operation the danger can be obviated by the use of de-fibrillators which enable electric shocks to be delivered through the chest wall and stimulate the heart muscle to beat in a co-ordinated manner once again.

Although, as already mentioned, small increases in body temperature do not appear to have any marked effects on cardiac output, apart from producing a raised heart rate, body temperatures in excess of 41°C (such as found in heat stroke) increase the likelihood of cardiac failure. This is not due so much to an episode, such as ventricular fibrillation, as to a dramatic reduction in cardiac output because of failure of the heart muscle to contract effectively.

Equality of temperature

A major role of the circulation in temperature regulation is in the distribution of heat throughout the body. There are different rates of heat production in various organs of the body, and without the circulatory system there could be very marked temperature differences between, for instance, the liver and the resting muscle in the back. Alternatively, in actively contracting muscle there could be a considerable temperature rise if the circulation of blood were to cease. An important role for the circulation is to prevent such inequalities; blood temperature is raised as it passes through an area of active heat production, and will tend to fall as it passes through regions that are less active. There will be a tendency to equality of temperature by this mixing process and as a result there is relatively little difference in the temperature of various organs of the body. Even though heat production in the liver is high the blood flow through this organ is also large and hence the temperature differences around the liver are very small as has been shown by measurements of temperature throughout the alimentary tract using the radio pill designed by Wolff[1].

Once the pill has left the stomach, on its journey along the alimentary tract (this takes an average of 24 hours) the temperature remains constant until the pill reaches the rectum where it will vary according to the activity of the subject as well as the exact position of the pill itself. Throughout the passage through the small intestine, and in the first part of the large intestine, no temperature changes are noted. This strongly suggests that the liver itself cannot have a temperature markedly different from other regions within the abdominal cavity because, for part of the journey, the pill will be in very close contact with the liver itself.

Temperature and blood vessels

The effects of temperature on blood vessels are complex owing to the fact that different vessels have a varying response to temperature and also to the complex innervation of the vessels and their distribution throughout the body. If isolated blood vessels are examined they do not appear to respond to temperature; there is no increase in diameter with heat, or a constriction in the cold[2, 3]. Similarly, if isolated vessels are perfused there is no change in the perfusation rate when the temperature of the perfusate is altered. On the other hand, some intact blood vessels within the body can show marked changes due to temperature. In discussing this it is first necessary to look at local temperature changes on blood vessels as measured with the plethysmograph. When the temperature within the plethysmograph is altered, blood flow is also changed; as the temperature decreases blood flow diminishes and as the temperature is increased blood flow goes up. This is shown in detail in Figure 6.2 where it is seen that over quite a wide temperature range at the lower end of the scale (from about 15 to 30°C) there is remarkably little increase in blood flow. From 30°C up to about 37°C a small increase in temperature is accompanied by a marked increase in blood flow[4, 5, 6, 7]. The blood vessels are very responsive to these temperatures which are in the range to which they

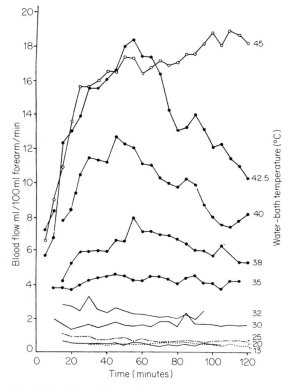

Fig. 6.2 The effect of temperature on forearm blood flow measured using a water bath plethysmograph (see also Fig. 6.8). Barcroft and Edholm.[5]

are most frequently exposed within the body itself. These changes, as will be seen later, are confined to the skin blood vessels and may be described as being direct effects of temperature on the vessels themselves because they are not influenced by denervation. If the nerve supplies to the blood vessels are divided by sympathectomy (removal of the sympathetic or autonomic nerve supply) the vessels still respond to temperature change. For example, in a patient with a sympathectomy the blood flow in the forearm will change, in the same way as in a normal subject, to variations in temperature within a plethysmograph. Blood vessels are not only affected by the temperature to which they are exposed but also by nerve impulses from vaso-motor nerves which may be either vaso-dilator or vaso-constrictor.

Direct and indirect effects of temperature

Direct temperature effects on blood flow are confined to the skin blood vessels and there is no effect of a rise in temperature on the blood vessels of the muscles. There may be some constrictor effect when the temperature is lowered although the main effect on muscle blood flow is probably due to changes in viscosity.

The indirect effect of temperature can be demonstrated in various ways. If the nerve supply is interrupted by the use of an anaesthetic (such as Novocaine) the effect on blood flow will depend on the degree of activity of the vaso-motor nerves. If, for example, the nerves are blocked by Novocaine and there is no change in blood flow, then it appears that either the vaso-dilator or vaso-constrictor impulses are balanced and that removing both of them has no effect upon the flow, or that the vessels are only moderately innervated.

If a nerve block is carried out to the fingers, a very marked increase in blood flow is recorded[8]. It may be concluded that the blood vessels of the skin in the fingers are richly supplied with vaso-constrictor nerves; this does not of course imply that there are no vaso-dilator impulses but that vaso-constriction appears to cover the whole range of blood flow. In the forearm, by way of contrast, there is little change in blood flow with a nerve block and this implies that there is very little vaso-constrictor tone[9]. Similar studies have been done on blood vessels in other body regions, particularly in the face, where it can be shown that there are varying degrees of vaso-constrictor tone in different regions. This has been studied in detail by Fox and Wyatt[10] whose results are summarized in Figure 6.3. The effect of vaso-motor nerve supply to the skin vessels can also be demonstrated by raising the temperature of the regulating centre in the brain. If, for example, the feet are immersed in hot water there will be an increased blood flow in the forearm. This effect is abolished by block of the forearm nerves including the sympathetic ones (Figure 6.4). The effect of heating by immersing the legs in hot water is to increase the blood flow in the forearm by reducing vaso-constrictor tone and possibly increasing the impulses from vaso-dilator nerves[11].

In sympathectomized limbs there is no effect on blood flow when the legs are immersed in hot water although they still respond to local temperature changes. It is curious, therefore, that isolated blood vessels do not appear to respond to temperature change. However, as it can be shown that these effects are confined to the skin vessels, and are not observed in muscle blood vessels, it seems possible that if the isolated vessels were taken from the skin and examined they might indeed show a response to temperature. This experiment has so far not been attempted.

The effects on the forearm blood flow of immersing the legs in hot water can also be abolished by a tourniquet on the legs. This prevents hot blood from the heated part of the legs reaching the temperature regulating centre in the brain. In this way it can be shown that the temperature effects are due to a vaso-motor reflex response. The degree of change in the forearm, due to the immersion of the legs in hot water, will depend on the temperature to which the forearm itself is exposed. If the blood flow is measured with the arm immersed in water at 20°C the resting flow will be as low as 1 ml/100 ml tissue/min and immersing the legs in hot water only has a very small effect; there will be a slight increase in blood flow but probably not more than 0.1–0.5 ml/100 ml tissue/min. If the forearm is exposed to a temperature of around 30°C then the response to leg heating will be much more marked and there may be a doubling of blood flow. If the local temperature is raised to

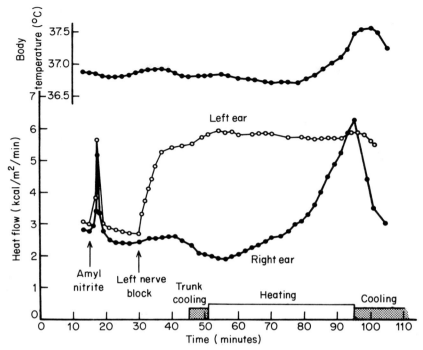

Fig. 6.3 The effect on heat flow from the left ear of blocking the greater auricular nerve. The effects of an inhalation of amyl nitrite before nerve block, and of body heating and cooling after nerve block, on the heat flows from both ears are also shown. (Reproduced with permission from the *Journal of Physiology*).[10]

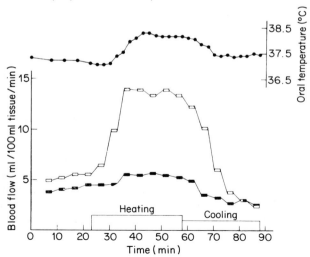

Fig. 6.4 The effect of cutaneous nerve block on the blood flow in the forearm during body heating and subsequent cooling. The right (■) forearm was anaesthetized and the left (☐) was the control.｜The oral temperature is also shown. (Reproduced with permission from the *Journal of Physiology*.)[9]

34°C there will be a further and very marked effect of leg heating as illustrated in Figure 6.5[12].

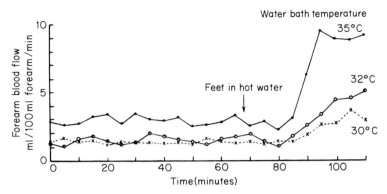

Fig. 6.5   Effect of local temperature on vascular response to heating the legs. Forearm blood flow was measured with the arm in water at 35°, 32°, and 30°C. At the arrow, the feet and legs were put into water at 43.5–44°C. The size of the subsequent vaso-dilation in the forearm is affected by the temperature of the water bath. Redrawn with permission from Barcroft and Edholm (1946)[5].

Evidence that changes of forearm blood flow are confined to skin

It has been stated that changes in forearm blood flow with leg heating are confined to the skin and are not associated or due to any change in muscle blood flow. The evidence for this is as follows. The blood flow in the skin can be estimated by techniques which abolish the flow. For example, iontophoresis of adrenalin can reduce skin blood flow to virtually nil. By measuring the flow in the forearm before and after this manoeuvre an estimate can be made of skin blood flow[13]. In the forearm, immersed in water at 34°C with normal body temperature, the skin blood flow estimated in this way is quite low and only accounts for some 10 per cent of the total forearm blood flow. In the iontophoresed arm there is no increase in blood flow with immersion of legs in hot water. Therefore, abolishing skin blood flow abolishes the response to leg heating. This in itself is evidence that blood flow is confined to the vessels of the skin and does not occur in muscles which are unaffected by the iontophoresis process.

An alternative technique is to use a nerve block of the skin blood vessels. If the cutaneous nerves of the forearm are blocked by the use of a local anaesthetic this also blocks the sympathetic or autonomic supplies to the blood vessels[8]. In such a nerve blocked forearm there is no change in flow when the legs are immersed in hot water (Fig. 6.6). Further evidence is provided by the effects of blocking the muscle nerves. Blocking the main nerve trunks to the forearm also results in blocking the muscle as well as the skin. However, the effect is to increase the forearm blood flow and this shows that the muscle blood vessels normally have a considerable vaso-constrictor

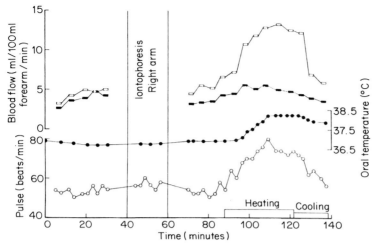

Fig. 6.6 Blood flows in both forearms (■ right, ▢ left) before and after adrenalin iontophoresis of the right forearm and showing the effect of body heating; also shown are pulse and oral temperature which remained relatively constant until the period of heating. (Reproduced with permission from the *Journal of Physiology*.)[8]

nerve supply. In such a muscle blocked forearm, there is no alteration in blood flow with the legs immersed in hot water.

Other workers have used different techniques to demonstrate the same point. Roddie, Shepherd and Whelan[14] have used the measurement of oxygen saturation from cutaneous and muscle blood vessels on the principle that if the blood flow through a part is increased, then the oxygen saturation will rise; if it is decreased the saturation will fall. It is therefore possible to follow alterations in the blood flow in different regions. By sampling directly the venous drainage from cutaneous and muscle blood vessels Roddie *et al.* were able to show that leg heating caused a very marked increase in the oxygen saturation of blood from the cutaneous veins, whereas there was no change in the muscle venous drainage. This confirms the effects described by nerve block.

These effects are illustrated in Figure 6.7. The blood flow in the limbs with the main trunks blocked increases in the resting state owing to the blockage of the vaso-constrictor supply to muscle blood vessels. However, local changes in temperature do not produce any further alteration. The nerve block has prevented any further dilation. Equally, if the forearm is cooled there is no change in blood flow; therefore local effects have been abolished by such nerve blocks. A few experiments have been done in which the brachial plexus has been blocked and others in which the sympathetic supply to the forearm has been interrupted by blocking the stellate ganglion in the neck; these procedures result in a very large increase in the forearm blood flow[15]. However, it is not yet clear where this increase occurs, whether it is entirely in the muscle blood vessels and why it should be a much larger increase with such a block than with a local block of say the ulnar or radial nerves. This is a

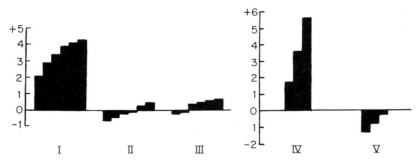

Fig. 6.7 The captions denote the condition of the test forearm. Ordinates: difference in test and control forearm blood flow in ml/100 ml forearm/min. (I) deep nerves blocked; (II) normal; (III) 'one' cutaneous nerve block; (IV) skin blanched, deep nerves blocked; (V) skin blanched. From Barcroft *et al.* (1943)[4].

difficult experiment to perform because of the inherent dangers hence there has been little work on this particular topic.

The conclusion from all of these various experiments is that the skin blood vessels have a fairly rich supply of vaso-motor nerves, not only in the fingers and hands but also in forearm skin and also in the cutaneous blood vessels of the face. These are activated by temperature changes in the vaso-motor centre. The muscle blood vessels do not have such a rich nerve supply and the nerves here appear to be mainly vaso-constrictor with an increased blood flow when their supply is blocked. Temperature affects the skin vessels but not those of muscle. There is no obvious structural difference between the two and the basis for this marked and significant difference is still not fully understood[16].

Techniques for studying blood flow

The main technique which has been used in the studies of Barcroft, Edholm and their colleagues is the volume plethysmograph. This is a simple and effective method of following blood flow but has some disadvantages. It can only be used for limb, hand or finger blood flow and it cannot be applied to the face.

The plethysmograph itself consists of a chamber which encloses the part to be studied; for example, the forearm. The chamber is sealed by rubber flanges at each end and filled with water. Changes in the volume of the enclosed part can be measured in a variety of ways for instance by measuring changes in the water level. The rate of blood flow is measured by occluding the arm above its point of entry into the plethysmograph using a pressure approximately equal to diastolic. The inflow of arterial blood is not therefore impeded so the volume of the arm within the plethysmograph will increase. The rate of increase as measured by the volume changes will be equal to the blood flowing into that part of the body which is contained in the chamber.

If the diastolic occluding pressure is maintained the venous pressure within the occluded part will eventually rise to be equal to or just above the diastolic pressure. Blood will then flow into the forearm and also return via the venous drainage; hence the use of the plethysmograph to measure blood flow has to be confined in practice to some ten seconds after the application of the occluding pressure.

Other techniques used for measuring blood flow include the clearance of radioactive Xenon. This method has been widely used in the USA and Scandinavia and has been progressively refined to be a very sensitive technique. The clearance rate depends on blood flow; the greater the flow the faster the rate of Xenon clearance. Essentially the same results are obtained with this method as with the plethysmograph.

Other more indirect methods include the skin plethysmograph. This depends on the reflection of light from the surface of the skin and is a function of the rate of skin blood flow. This method has also confirmed the previously described effects and may be used effectively on the face where it can be applied locally to assess blood flow changes qualitatively. It is not simple to make the skin plethysmograph into a quantitative instrument.

The use of skin temperature has also been used as an index of skin blood flow. This is a fairly crude technique but is useful for indicating changes in skin blood flow or, using deep tissue probes, changes in regions deeper than the skin. When skin temperature is used it is necessary to remember that small changes of blood flow with cold skin will produce relatively large changes in the skin temperature. As skin temperature approaches 37°C the effect, even of large changes in skin blood flow, are not mirrored by large changes in skin temperature. At a skin temperature of say 37°C no change in blood flow can be recorded because the temperature of the blood reaching the skin is itself at 37°C. Skin temperature is very useful as a qualitative indicator of cutaneous blood flow but it is difficult to use for quantitative measurement.

If the vaso-motor supply is blocked off so that reflex effects are abolished, for instance in one arm, and the blood flow is examined in the two arms, an estimate can be made of the proportion of total blood flow contributed by skin.

The results of these various measurement methods have shown that skin blood flow varies over a wide range in most parts of the body, being greatest in the fingers and toes and in parts of the face. The absolute level of flow in any particular area of the skin depends upon the richness of the cutaneous blood vessels as well as on the local environmental temperatures which determine the degree of vaso-constriction or dilation. The results show that blood flow in the extremities is mainly, if not entirely, controlled by vaso-constrictor nerve fibres. A nerve block in a finger results in an increase in blood flow to the maximum possible. A permanent block achieved by sympathectomy may be carried out for clinical conditions which present with abnormal vascular states, for instance in Raynaud's disease. Sympathectomy has also been used for the treatment of excessive sweating which may be very distressing to patients; denervating the sweat glands as well as the blood vessels resolves this condition. Such denervated or sympathectomized regions are very useful for studying vaso-motor control mechanisms and they allow an examination of the effects of environmental temperatures.

Over the rest of the body, the limbs and trunk, the vaso-motor supply appears to be both vaso-dilator and vaso-constrictor. Blocking the nerve supply to the skin of the forearm does not necessarily result in a large increase in blood flow. The face, including the lips and forehead, appears rather variable with a balance between vaso-dilator and constrictor nerves. The blood flow is in general high in the skin of the face and a nerve block usually results in an increased blood flow, indicating that there is a considerable vaso-constrictor nerve supply[17]. In the extremities there are shunts between arterioles and venules; these arterio-venous shunts are richly supplied with vaso-motor nerve endings. Under certain conditions, particularly in the cold,

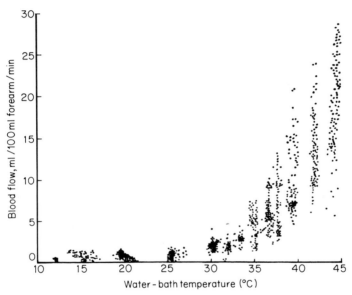

Fig. 6.8   Forearm blood flow plotted against water bath temperature. The forearm and hand were kept in water at a particular temperature for 2 hours. Blood flow declines with decrease of temperature. Reproduced with permission from Burton and Edholm (1955)[12].

these open to produce a large blood flow through the region by-passing the capillary blood vessels. The tips of the fingers are particularly rich in these shunts which diminish gradually along the fingers. They are also found on the palmar surface of the hand, in the toes and in the dorsal surfaces of the feet, around the knees and elbows and also on the lips and cheeks. There is no evidence of arterio-venous shunts being found in the blood vessels supplying the muscles. Figure 6.8 shows the changes in blood flow in the forearm and in the hands when they are immersed in water at various temperatures. The flow is very low when the water temperature is below 15-25°C. At 30°C there is some increase and this is very marked as temperatures rise up to 45°C which was the highest temperature used in these studies. The question arises as to where this increased blood flow is and how far it is due directly to the effect of temperature and how much to the modification of the vaso-motor nerve supply. This was easily determined by measuring the temperature effect on blood flow in sympathectomized limbs. There was little difference in the blood flow through the hand and forearm immersed in water at various temperatures between the sympathectomized or the intact limb. The vaso-dilator and constrictor nerve supply does not appear to have affected this local temperature response. However, it is still uncertain from such an experiment whether the increased blood flow is in the skin or in the muscles or both.

The striking difference between skin and muscle blood flow is still not understood. The vessels in the muscle respond very rapidly to other stimuli including the metabolic products of muscular contraction. The blood flow through the muscle appears to be unaffected by sympathectomy and increases with muscular contraction to the same degree as before the operation.

If body temperature as a whole is raised there is a marked increase in blood flow in the periphery even if the local skin temperature remains constant. This was shown by experiments where the blood flow was measured in the forearm immersed in water at 35°C. When body temperature was raised by immersion, in water which was gradually warmed up to 41°C, forearm blood flow increased dramatically to several times its resting level even though local skin temperature remained constant.

These temperature effects take place in the skin and are dependent both on local temperature and that of the regulating centre, that is the deep body temperature. The threshold body temperature at which vaso-dilation occurs is 37.4°C, irrespective of whether the temperature is rising or falling[18].

The increased blood flow produced by a rise of body temperature is influenced by the skin temperature, the lower this is the smaller the blood flow[17, 18]. There are two vascular adjustments involved in the increased skin blood flow consequent on body heating[18, 19]; arterio-venous anastomoses open and so do the superficial veins. These changes occur over a wide range of blood flow and in absolute terms this can be measured in millilitres of blood per 100 g of skin per minute. These calculations have been based upon dissection of the forearm into its component parts so that the quantity of skin may be measured. The flow of blood in the skin can be as low as 0.1 ml of blood flowing through 100 g of skin per minute or as high as 100 ml of blood per 100 g of skin per minute. Such large ranges are only found in the fingers and possibly parts of the face; over the rest of the body values from about 0.4–0.5 ml per 100 g of skin per minute to 30–40 ml are found.

As already mentioned, in the cold the arterio-venous shunts are open. This phenomenon of cold vaso-dilation was first described by Sir Thomas Lewis when he referred to a 'hunting' phenomenon. He used this term to describe the skin temperature changes that occur when a finger is immersed in iced water. There is an immediate and rapid fall to nearly the level of the iced water temperature. After approximately 4–5 minutes an abrupt rise of as much as 8–10°C occurs but does not persist. After a further 8–10 minutes a fall is observed but not to as low a temperature as before, and there are succeeding waves of dilation and constriction. It was this alternating rise and fall in skin temperature which Lewis described as the 'hunting' reaction.

Accompanying these changes there is considerable discomfort when the finger is first put into iced water; this gradually develops into pain which becomes increasingly severe. Then, rather abruptly, the intense pain is succeeded by a feeling of warmth and the subject will be quite convinced that the water temperature has been raised, even though it has remained unchanged. This is the period when the arterio-venous shunts open and the blood flow through the finger is greatly increased, accompanied by a rise in skin temperature. After a time, there is a recurrence of discomfort and pain again followed by a feeling of relief as the temperature and the blood flow increases. These periods of constriction and dilation have been examined in many circumstances but it may be said that the mechanism is still unclear. Cold vasodilation can occur, however, following sympathectomy which does not have any apparent effect upon this response. On the other hand, there is evidence which Lewis, himself, produced that the response depends upon the integrity of the somatic nervous system. If for instance the nerve supply to the hand has been destroyed and it is completely anaesthetic, cold vaso-dilation is very much harder to demonstrate and is usually absent. This led Lewis to believe that the response was due to an axon reflex from the sensory nerves in the branch going directly to the vessels to cause the dilation. However, Greenfield, Shepherd and Whelan[20, 21, 22] produced evidence showing it to be improbable that the effect was due to such an axon reflex; they examined patients with peripheral nerve lesions and were able to show a cold vaso-dilation, even though it might be smaller than in normal hands or fingers. Other theories have been put forward including one by Keatinge[3], who suggested that the blood vessels cool and eventually become paralysed and unable to respond to any substance including adrenalin or noradrenalin. He has also shown, by examining isolated blood vessels, that at about 10°C or lower, the response to drugs is either abolished or greatly reduced. Keatinge supposes that when the blood vessels become paralysed the wall relaxes and the flow is restored; as the vessels themselves warm up they again become responsive not only to circulating drugs such as adrenalin but also temperature; they will produce a cycle of constriction and paralysis which accounts for the 'hunting' reaction. This 'hunting' has been examined in relation to acclimatization or adaptation of the fingers and hands to cold. Yoshimura and his colleagues in Japan[23], examined people living in very cold regions and were able to show that those whose hands were habitually exposed to cold exhibited differences in the cold vaso-dilation response. These differences are first that the initial constriction is less marked than in the normal finger and that the blood flow does not entirely cease. Secondly, the cold vaso-dilation begins earlier and the consequence of both of these effects is that the subject feels little or no pain although there may still be some discomfort. Thirdly, the dilation appears to persist for longer than in the unadapted hand.

Such adaptations have been considered as the basis for the way in which people can continue working with their hands in very cold water. Examples are the salmon fishermen of Gaspé Peninsula who have been examined by Le Blanc[24] who has shown that people habituated to this very cold water exhibit the cold vaso-dilation phenomenon described above. This effect persists for some time after subjects cease exposing their hands to the cold but does wear off after a period of approximately a year. This adaptation has also been observed in the Eskimo and in Lapps[25, 26, 27] who are accustomed to having their hands exposed to very cold conditions. The cold vaso-dilation in itself may be regarded as an effective protection of the hands against local cold injury. This is not peculiar to man and it was demonstrated, also by Lewis, and his associates many years ago, that this could be seen in the ears of rabbits and the feet of birds, such as gulls in the arctic. Another aspect of this effect which has been examined is the state of body temperature. It is much harder to demonstrate cold vaso-dilation in the hands of a subject who is generally cold. If his body temperature has fallen, or if his skin temperature generally is low and he is shivering, then the local cold vaso-dilation response is markedly diminished. Fox and Wyatt[10] have also shown this effect can be demonstrated in the knees, elbows and face. All of these regions have arterio-venous shunts and the relationship between the shunts and the response of the vessels seems now to be quite unequivocal. Attempts have been made to find such shunts elsewhere in the skin of the body but none have so far been demonstrated.

There is still much to be learned about the control of the cutaneous blood vessels. This is evident from a study of the thermograms shown in Plates 1–16. Skin temperature is influenced by environmental conditions, but the two dominating factors are cutaneous blood flow and heat transfer from underlying tissues and organs. In resting conditions, the blood flow through the skin is the most important regulator of skin temperature. What the thermograms have revealed is the wide range of skin temperature which appears to be normal (Plates 1 and 5). Furthermore, there are frequently changes of temperature for no immediately obvious reason (Plate 11). Sometimes, as in sleep, there are rhythmic changes, with overall skin temperature rising and falling. Sometimes they are apparently random unsymmetric alterations.

Such variation and variability in skin temperature could be due to local changes in heat production. Muscular activity can provide the means for varying heat production, but the thermographic technique shows that there can be fluctuations of skin temperature in a state of complete rest.

It can be concluded that to a large extent the skin temperature patterns and their mutability are a reflection of alterations in skin blood flow. Although there is evidence for rhythmic discharges from the vaso-motor centre these could not account for the wide range of skin temperature or the localization of changes. Our present understanding of the vaso-motor control of the cutaneous circulation is inadequate to explain the thermographic findings. There is clearly a fruitful field for research using this technique.

References

1. Wolff, H.S. (1961). The radio pill. *New Scientist* **12**, 419.
2. Keatinge, W.R. (1957). The effect of general chilling on the vasodilator response to cold. *Journal of Physiology* **139**, 497–507.
3. Keatinge, W.R. (1958). The effect of low temperature on the responses of arteries to constrictor drugs. *Journal of Physiology* **142**, 395–405.

4.   Barcroft, H. and Edholm, O.G. (1943). The effect of temperature on blood flow and deep temperature in the human forearm. *Journal of Physiology* **102,** 5–10.

5.   Barcroft, H. and Edholm, O.G. (1946). Temperature and blood flow in the human forearm. *Journal of Physiology* **104,** 366–76.

6.   Allwood, M.J. and Burry, H.S. (1954). The effect of local temperature on blood flow in the human foot. *Journal of Physiology* **124,** 345–57.

7.   Johnson, J.M., Brengelmann, G.L. and Rowell, L.B. (1976). Interactions between local and reflex influences on human forearm skin blood flow. *Journal of Applied Physiology* **41,** 826–31.

8.   Edholm, O.G., Fox, R.H. and Macpherson, R.K. (1956). The effect of body heating on the circulation in skin and muscle. *Journal of Physiology* **134,** 612–19.

9.   Edholm, O.G., Fox, R.H. and Macpherson, R.K. (1957). Vasomotor control of the cutaneous blood vessels in the human forearm. *Journal of Physiology* **139,** 455–65.

10.   Fox, R.H. and Wyatt, H.T. (1962). Cold induced vasodilatation in various areas of the body surface of man. *Journal of Physiology* **162,** 289–97.

11.   Cooper, K.E., Edholm, O.G. and Mottram, R.F. (1955). The blood flow in skin and muscle of the human forearm. *Journal of Physiology* **128,** 258–67.

12.   Burton, A.C. and Edholm, O.G. (1955). *Man in a Cold Environment.* Edward Arnold, London.

13.   Barcroft, H., Bonnar, W.McK. and Edholm, O.G. (1947). Reflex vasodilatation in human skeletal muscle in response to heating the body. *Journal of Physiology* **106,** 271–8.

14.   Roddie, I.C., Shepherd, J.T. and Whelan, R.F. (1956). The effect of heating the legs and of posture on the blood flow through the muscle and skin of the human forearm. *Journal of Physiology* **132,** 47P.

15.   Barcroft, H., Edholm, O.G., Foster, C.A., Fox, R.H. and Macpherson, R.K. (1956). The effect of nerve block on forearm blood flow. *Journal of Physiology* **132,** 16–17P.

16.   Fox, R.H. and Edholm, O.G. (1963). Nervous control of the cutaneous circulation. *British Medical Bulletin* **19,** No. 2, 110–14.

17.   Fox, R.H., Goldsmith, R. and Kidd, D.J. (1962). Cutaneous vasomotor control in the human head, neck and upper chest. *Journal of Physiology* **161,** 298–312.

18.   Cabanac, M. and Massonet, B. (1977). Thermoregulatory responses as a function of core temperature in humans. *Journal of Physiology* **265,** 587–96.

19.   Kerslake, D.McK. (1972). *The Stress of Hot Environments.* Cambridge University Press, London.

20.   Greenfield, A.D.M., Shepherd, J.T. and Whelan, R.F. (1950). The average internal temperature of fingers immersed in cold water. *Clinical Science* **9,** 349–61.

21.   Greenfield, A.D.M., Shepherd, J.T. and Whelan, R.F. (1952). Circulatory response to cold in fingers infiltrated with anaesthetic solution. *Journal of Applied Physiology* **4,** 785–8.

22.   Greenfield, A.D.M., Shepherd, J.T. and Whelan, R.F. (1951). The part played by the nervous system in the response to cold of the circulation through the finger tips. *Clinical Science* **10,** 347–60.

23.   Yoshimura, H. and Iida, T. (1952). Factors governing the individual difference of the reactivity of the resistance against frostbite. *Japanese Journal of Physiology* **1,** 177–84.

24.   Le Blanc, J. (1956). Impairment of manual dexterity in the cold. *Journal of Applied Physiology* **9,** 62–4.

25.   Folkow, B., Fox, R.H., Krog, J., Odelram, H. and Thoren, O. (1960). An analysis of the vascular response to intense cooling. *Acta physiologica Scandinavica* **50,** Suppl. 175, 49–50.

26. Fox, R.H., Krog, J., Odelram, H. and Thoren, O. (1963). Studies on the reactions of the cutaneous vessels to cold exposure. *Acta physiologica Scandinavica* **58,** 342-54.
27. Krog, J., Folkow, B., Fox, R.H. and Lange Andersen, K. (1960). Hand circulation in the cold of Lapps and North Norwegian fishermen. *Journal of Applied Physiology* **15,** 654-8.

7

Man is a Tropical Animal

Thermal characteristics of man

A distinguished biologist, Peter Scholander, coined the phrase—'man is a tropical animal'—many years ago. He was emphasizing that man almost certainly originated in East Africa in a hot, possibly rather dry, savannah environment, conditions for which man is well adapted. He is extremely well-endowed with sweat glands and has more than any other mammal; because of the ability to sweat at a high rate man can maintain body temperature in hot climates without difficulty. There is a price to pay, requiring the replacement of the water and salt lost, but in many cases this is not a serious problem. Man is also almost hairless; this is not necessarily a great advantage in any climate but it is less of a disadvantage in hot climates than it is in cool or cold regions. Certainly for a highly sweating animal, a thick fur coat would be incompatible. If man were to live naked and maintain a body temperature close to 37°C he must have a climate where the temperature is of the order of 28–30°C. Otherwise, he has to have artificial means of insulation from the environment or suffer the discomforts and physiological cost of increasing heat production by shivering, or by active muscular movements. For all of these and many other reasons, man has the characteristics of a tropical animal.

Another characteristic, which has already been mentioned, is that the metabolic rate of man in the resting nude state begins to rise when the environmental temperature drops below about 28°C; this is comparable with findings for other tropical animals.

Scholander et al.[1] examined the metabolic response of animals to cold, and showed that the most important feature was the degree of insulation provided by fur. Animals adapted for life in sub-arctic conditions had such a thick coating of fur that they could not only withstand very cold conditions but could do so without increasing their metabolic rate. The critical temperature for such animals could be as low as −40°C as in the case of the arctic fox. Scholander[2] measured the sub-clothing temperature of the Eskimo living in temperatures down to −40°C and showed that it was virtually the same as in men living in temperate or even warm environments. He claimed that the Eskimo is not so much adapted to cold but he has contrived his environment to maintain a tropical climate beneath his clothing. Similar observations have been made since 1955 by members of Antarctic expeditions who were able to make measurements of microenvironmental temperatures beneath clothing and next to the skin[3] without disturbing or constraining the subject. Over a period of a year in the Antarctic the average sub-clothing temperature was approximately 33°C and similar to that recorded in people living in a temperate climate. This is approximately equivalent to a mean skin temperature of 34°C which is the usual level measured in conditions of thermal comfort. Man adapts himself by appropriate clothing to maintain such a skin temperature wherever he is living, and it turns out that this is similar to that which would be found under tropical conditions. These findings confirm Scholander's view that 'man is a tropical animal' and emphasize the importance of looking at the way in which he adapts physiologically to different thermal environments.

We can take, as the starting point, the fact that man remains a tropical animal whatever conditions he is living in but there are differences between people who are living in cool climates and those who live in tropical conditions in their response to heat. Compared with people who habitually live in cold climates, those in hot regions are at an advantage when it comes to working in the heat. The newcomer to the tropics who arrives from cool regions is at first very uncomfortable. It is difficult for him to do any physical work and impossible for him to do hard work without the risk of collapsing from a raised body temperature due to the onset of 'sweat exhaustion'. This means the fall in sweat rate which occurs in people who are exposed to heat and have to maintain a high evaporation rate in order to keep the body temperature under control. In such conditions exhaustion of the sweat glands can occur and they are described as being 'fatigued'. Heat collapse is rare amongst people who habitually live in hot countries because they know how to live in the heat, they wear appropriate clothing and work early in the cool of the morning or in the evening when the sun has gone down and temperatures have fallen. It is only in unusual circumstances that people will work hard in the midday hours in a hot country.

Apart from such behavioural adaptations there are differences on exposure to heat in heart rate, sweat rate and the rise of body temperature. The unadapted individual will have a lower sweat rate, a higher body temperature and a higher heart rate than the indigenous subject doing exactly the same work under the same conditions. However, over a period of time, which is measured in terms of two to three weeks only, the newcomer, provided he is doing some work in the heat, will adapt and have a performance which is not greatly different from that of the indigenous subject. Adaptation, or acclimatization to heat, is easy to demonstrate; the physiological changes are quite unequivocal and they have been found in everybody who has been examined, in men and women, in the young and old, in the fat and thin and in people who are athletic and those who are sedentary.

What about the other way round; can changes be demonstrated in people who move from a hot to a cold climate? In a sense, one set of changes can be described; when people who are adapted for life in the tropics, come to live in a cold climate they will lose the acclimatization to heat, and on return to the tropics will need to become reacclimatized. Changes which occur in people who go from the UK and live and work in the sub-arctic, northern Canada, or to the Antarctic are small and difficult to demonstrate. There is no obvious change in any general physiological characteristic although there may be some changes in thyroid activity and vascular responses. There does not appear to be any change in the mean skin temperature for thermal comfort and there is no clear evidence of changes in metabolic rate. The negative findings are in contrast with laboratory work on animals as diverse as the rat, the rabbit, the guinea pig and the monkey, all of whom have marked changes associated with adaptation to cold that include metabolic changes and the phenomenon of non-shivering thermogenesis, an increase in metabolism not due to shivering in muscles. Such non-shivering thermogenesis has not been unequivocally demonstrated in man although there is some suggestive evidence for it.

It is fairly clear that man's specific adaptation to cold is behavioural. He learns to dress to maintain his accustomed skin temperature, and to behave in the cold to keep up his body temperature, by muscular work. He learns to avoid cooling from wind by sheltering from it, and so on. All of these adaptations are evident if one examines the behaviour of people living in cold places; they tolerate the cold conditions and it is this behavioural aspect which is really the means that they have for protecting themselves against the effects of cold. This point can be illustrated by examining the incidence of cold injury. In order to see cases of frost-bite one should not go to the north of Canada or northern Scandinavia but to those large cities of the world which have cold winters such as Chicago or New York. Here one finds people who are likely to drink too much and who may collapse in the streets and be at risk of severe cold exposure including frost bite. In a Canadian survey of the incidence of frost-bite, cases were found only in the large cities such as Ottawa and scarcely any were noted in people living north of Winnipeg. Behaviour is clearly the key to man's adaptation to cold.

Acclimatization

The term 'acclimatization' is used to describe the changes that take place when an individual moves from a cold to a hot climate or vice versa. There are clear and precise physiological changes which have been extensively investigated in people moving from cold to hot regions, and they have been confirmed by studies in climatic chambers in which a wide range of environmental conditions can be produced.

Acclimatization can be described by following the changes that occur in a group of men exposed in a climatic chamber day after day.

Such studies were carried out in London at the Queens Square Laboratories of the Medical Research Council for the Royal Naval Personnel Research Committee where, in a fairly crude climatic chamber, many experiments were carried out between 1942 and 1946. A routine was prepared based upon the predicted activity pattern of ratings serving guns with ammunition during a naval engagement. These men would work and rest alternately in half-hour periods for up to 4 hours. This routine was used in climatic chamber experiments because it was then possible to compare the results of subsequent investigations with the original work at Queens Square. These studies were extended after the war, by the Medical Research Council in association with the Royal Navy at Singapore where a whole series of experiments were performed with naval personnel. The results were summarized by MacPherson in 1960[4]. Further studies in the climatic chambers of the MRC laboratories at Hampstead were made between 1958 and 1970 and the work patterns again followed the same routine as at Queens Square. The results of the Queens Square investigations led to the development of the predicted 4-hour sweat rate (P4SR) index (see Chapter 9).

The climate to which subjects were exposed in the chambers was fixed at 40°C dry bulb and 32°C wet bulb with an air movement of 0.2 m/s; the walls of the climatic chamber were also kept at 40°C.

Subjects carried out a standardized work task, stepping up and down on a stool 12 times a minute; the energy expenditure averaged 750 kJ/m²/h. After working for half an hour the subjects rested, sitting on the stool; they resumed

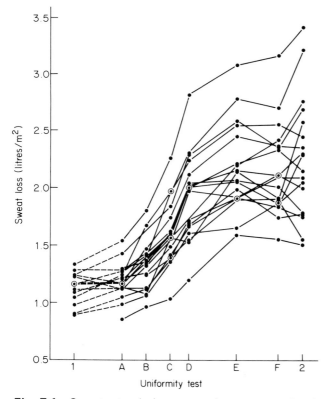

Fig. 7.1 Sweat rates during successive exposures in a hot room. Eighteen subjects spent 4 hours, on five days in the week, in the hot room. Conditions varied slightly from day to day but, on the occasions illustrated, the environment was identical. On Day A, at the beginning of the experiment, the volume of sweat produced in 4 hours ranged from 0.8 to 1.55 litre/m². On the final day (labelled 2) the sweat volume was from 1.5 to 3.4 litres/m². The sweat rate had approximately doubled in each subject; the subject with the lowest sweat rate remained lowest throughout the experiment and the highest remained the highest.[8]

stepping after half an hour. The subjects were weighed before entering the chamber, heart rate and body temperature were measured at the beginning and end of each work period; measurements were made of any urine or vomit. Sweat rate was determined from the differences in weight of the subjects at the beginning and end of the period, and corrected for any water drunk or urine passed.

On the first day, many subjects were distressed and some collapsed with a heat syncope resembling fainting; others experienced nausea and vomiting. During the 4 hours, body temperature would rise to over 39°C but subjects were removed if it exceeded 39.5°C. Heart rate increased and at the end of the last work period would average approximately 180 beats a minute; sweat rate was approximately 2 litres (or 0.8–1.5 litre/m²) for the 4-hour period.

On subsequent days body temperature and heart rate did not increase so much but the sweat rate rose. By the end of three weeks with five days exposure each week, sweat rates had on average doubled; the body temperature rise did not exceed 37.6°C and the heart rate was approximately 130 beats a minute. These were substantial physiological changes and occurred in men who were studied in the UK and in the USA. Many of these experiments were done in the winter and in the USA the outdoor temperatures at the time were well below 0°C. However, in spite of the exposure to these cool or cold conditions, the same physiological changes took place and were independent of the climatic conditions outside the chamber. These results are illustrated in Figures 7.1, 7.2 and 7.3.

Differences in acclimatization status

People living in hot countries were also examined. When comparisons were

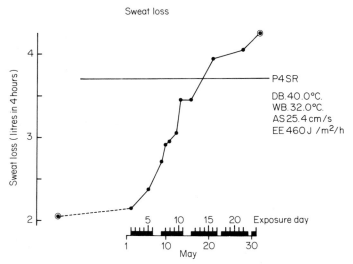

Fig. 7.2   The mean results of the sweat rates shown in Figure 7.1 given in litres/4 hrs: DB = dry bulb temp; WB = wet bulb temp; AS = airspeed; EE = energy expenditure.[8]

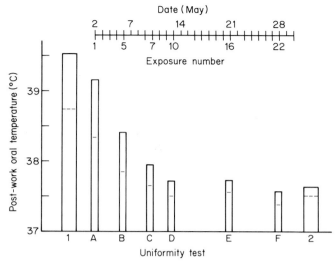

Fig. 7.3 The oral temperature at the end of the 4-hour exposure. Mean results in the 18 subjects of Figure 7.1. After the tenth day of exposure the rise in body temperature remains approximately constant.[8]

made between observations in Singapore (with a hot, humid climate) and in the UK the results showed that the acclimatization process was the same in the two countries, although the Singapore subjects (also naval ratings) could be said to have had 3–4 days of acclimatization to start with[5]. Their sweat rates were higher and body temperature and heart rate responses were rather less than those of the UK subjects entering the hot chamber for the first time. Otherwise, the time course of the changes was similar. Nigerians[6] have a degree of acclimatization but they can be further acclimatized by exposure in a climatic chamber at the levels of temperature used in the studies in the UK and USA; when this is done the pattern of change seems to be similar in many respects.

Comparison of British and Indian soldiers
A further experiment was carried out when members of the Indian army were flown to England at the end of the hot season (in late September) after military training in the hot weather of Lucknow. On arrival in England, they were tested in a climatic chamber in exactly the same way as the British soldiers had been. The Indians had sweat rates which were higher than those of the 'novice' English soldiers but not as high as fully acclimatized British soldiers. The body temperature and heart rate responses indicated a degree of acclimatization, but this was not complete. The Indian soldiers remained in England during the autumn and winter which was a severe one. They were not further exposed in the climatic chamber. The men, who were training at camps in England, worked out of doors in the cold throughout the winter. They were re-tested in the climatic chambers after being in England for four months. The acclimatization that they had on arrival had gone; their sweat

rates were low, even lower than those of British soldiers at the same stage, and they had body temperature and heart rate responses similar to those of people who were completely unacclimatized to heat. The soldiers were subsequently exposed to the acclimatization procedure in the climatic chamber. They never achieved the same high levels of sweat rate as the acclimatized British soldiers; their sweat rate increased to approximately the same level as when they had first arrived from India. The same was true of the heart rate and body temperature response[7] (see Figs 7.4 and 7.5).

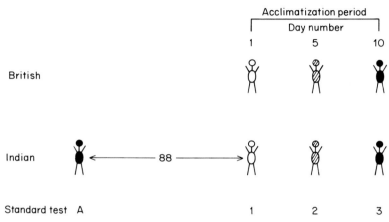

Fig. 7.4   Plan of the experiment comparing Indian and British soldiers. A = hot chamber test after arrival from India. 1 = test of unacclimatized British and Indian soldiers after winter spent in England. 2 = state following one week acclimatization, both groups partially acclimatized. 3 = after two weeks in hot room, both groups fully acclimatized.

Comparison of natural and artificial acclimatization

Another of the problems of acclimatization to heat is to establish whether acclimatization acquired in a climatic chamber is similar or identical to that acquired by living in a hot country. This has been tested in two large-scale experiments carried out in collaboration with the British Army[8]. In the first of these the subjects, who numbered 120, were exposed to a standard test within the climatic chamber to measure each individual's sweat rate, body temperature and heart rate response. The subjects were weighed and their heights were measured. On the basis of these parameters they were divided into three matched groups. The first known as the control group, was sent to Scotland at the end of March to continue their training in a cool climate. The second, called the 'naturally acclimatized group' was sent to Aden (latitude 10° north), at that time a British colony, where they were to train in weather which would become increasingly hot; the third termed the 'artificially acclimatized group' remained in London and were exposed daily in the climatic chambers at Hampstead to standard hot conditions (40°C dry bulb, 32°C wet bulb and an air movement of approximately 0.2 m/s).

Treatment of the three groups continued for a period of five weeks. The

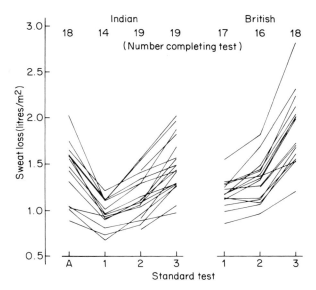

Fig. 7.5 Sweat rates in Indian and British soldiers at various stages (see Fig. 7.4). British soldiers reached higher sweat rates than did the Indians after acclimatization.

subjects in the climatic chambers spent 4 hours daily carrying out a standard work/rest routine. Their body temperature, heart and sweat rates were continuously monitored.

At the end of the treatment period the control group and the naturally acclimatized group were brought back to Hampstead by air and all three groups were tested in the climatic chambers. This showed that the artificially and naturally acclimatized groups had fairly similar responses in the chamber. The control group had significantly lower sweat rates and higher heart rates and body temperatures at the end of the 4 hours compared with the other two groups. The results of the treatment had been to separate the control group from the other two. Immediately after the tests were completed all three groups were flown out to Aden where an arduous 12-day exercise was carried out, during which the men's military performance was assessed. There were three platoons with three squads each and in each platoon there was a squad from each of the treatment groups. In this way, it was hoped that the possible effects of the standards of the different platoon commanders would not affect the results. In the event, one of the three platoon commanders proved to be outstandingly good and the performance of his platoon was significantly better than that of the other two. Nevertheless, owing to the design of this experiment it was possible to assess the performance of the members of the three treatment groups.

During the 12-day exercise, measurements were made of water and food intake and of sweat loss assessed by weighing; body temperature was measured at hourly intervals together with heart rate.

Finally, full accounts were kept of any heat collapse or illness in the field. Separate assessments were made by experienced army officers of the military performance of the three platoons and the different squads within the platoons. At the end of the exercise in Aden all of the subjects, together with the army observers and physiological team, were flown back to England and all subjects were tested in the climatic chambers. Thereafter, the rate at which acclimatization fell off was assessed by testing all individual subjects once a week over a period of six weeks.

The results showed that the three groups behaved differently in Aden in that the naturally acclimatized men had the best military performance throughout the 12-day exercise; the control group performed worst for the first half of the exercise with the artificially acclimatized men having an intermediate performance. In the second half of the experiment when there was virtually no difference between the artificially acclimatized and control groups, the naturally acclimatized men performed best. The food intake in all three groups fell steadily throughout the exercise and reached very low levels on some days. The initial intake was of the order of 12 500 J (3000 kcal) per day and this fell to just above 4200 J (1000 kcal) per day and there was a substantial drop in body weight. There was no difference between the three groups. All subjects had a high water intake averaging 8 litres per man per day; one subject drank 15 litres in one day. The intake was constantly slightly lower amongst the control subjects.

Initially the artificially acclimatized group had the highest sweat rate, the naturally acclimatized had rates which were close to them and the control subjects had the lowest rate. During the course of the trial, the sweat rates became practically equal for the three groups, implying that the control group were becoming acclimatized to heat. The highest temperatures were recorded in the control group and the lowest in the artificially acclimatized with the naturally acclimatized only slightly higher. These differences diminished during the trial and by the end there was little to choose between the groups. Heart rate changes followed a similar course to body temperature.

The military performance of the naturally acclimatized subjects was the best throughout. The artificially acclimatized men were not so good at any time and were even worse than the control group during the last few days of the exercise. It was then realized that apart from the different environmental experience of the three groups, there had been marked contrasts in military training. The men who had the artificial acclimatization procedure at Hampstead had little time for military training after travelling to and from the laboratories and their barracks. Both the other groups had energetic training in Scotland and in Aden. It seems probable that this lack of training prejudiced the military performance of the artificially acclimatized group in Aden.

Finally, one of the factors which had not been examined in the UK was intelligence level. This was measured in Aden and there were significant differences between the groups. The naturally acclimatized men had the highest level of intelligence while the artificially acclimatized had the lowest level which could have affected their performance.

The differences in military performance between the three groups may very well not have been related to their treatment regarding exposure to heat. On the other hand, the physiological measurements certainly showed some differences between the three groups although these diminished as the trial proceeded. There was one other measure, namely the number of heat casualties; in the first half of the trial the artificially and naturally acclimatized groups had about the same number of casualties. In the second half the naturally acclimatized group had a lower rate of casualties than either the artificially acclimatized or control groups. The conclusions were that the artificially acclimatization did confer a benefit but the effects of natural acclimatization appear to be more effective.

The final set of observations on these subjects was concerned with the decline of acclimatization that had been acquired and this showed itself in the first trial on arrival back from Aden when there was little difference between the three groups in their responses in the hot chamber. They were then tested at weekly intervals and it appeared that acclimatization decayed over a period of about nine weeks. This suggests that the rate of decay was fairly similar to the rate of onset of acclimatization itself; the decay being perhaps slightly slower[9].

There were many questions which arose during the course of this trial, specifically the problem of heat illness. It was hard to predict which subjects might become heat casualties as there was no clear relationship between performance in the climatic chamber and performance in the field. Cardiovascular stability tests were not useful as predictors either. The only relationship found was between subjects' body weight and liability to heat illness. The heavier the individual, which in most respects meant the fat subject, the more likely he was to suffer from heat illness.

In view of the difficulty experienced in this trial it was decided to carry out a second major experiment with the field work in the hot climate of Aden. Throughout the trials, both in London, Scotland and Aden, climatic conditions were regularly measured. In Aden the climate is severe, with high daily temperatures of the order of 36-40°C and high relative humidity (of the order of 50-60 per cent during the day) rising to as much as 100 per cent at night with occasionally very strong winds. There was considerable radiation reflected from the surrounding desert surfaces and rocky hills where the exercises were carried out.

The second experiment[10] was designed to avoid the difficulties concerning physical fitness and training. It was decided to choose men of the Parachute Brigade who need to have an exceptionally high level of physical fitness to be accepted and who maintain a very high standard of training. Instead of artificially acclimatizing men in chambers it was decided to use naturally acclimatized subjects. The Parachute Brigade at that time maintained a company in Bahrain in the Persian Gulf, and another company stationed in the UK. These companies were due to replace each other after one year's service in Bahrain.

The trial was carried out in three phases. In the first phase, the unacclimatized (to heat) platoon was tested in England before any of the members had been exposed to heat. In the second phase, the platoon from England was joined by a platoon from Bahrain, both groups flying to Aden at the same time. This phase was carried out in Aden, whilst for the third phase both the groups were brought back to England. Each of these three phases was as similar as possible in terms of military activity; the only difference being climatic. In each phase, lasting 12 days, there were three stages. During the first four days, termed 'hard', the men carried out arduous physical training; the second four days were termed 'soft', when the men had a rest from any military training and the third period of four days was a 'hard' period with heavy physical work. In each phase the timetable from the first to the twelfth day was made as similar as possible. In Aden the exercise which was carried out in the 'hard' days was on terrain that had been carefully mapped beforehand and where the activities were matched to those that could be carried out in England at Aldershot. The measurements made were similar to those in the first trial, that is food and water intake, urine output and faecal mass were all measured daily. From these figures and from changes in body weights (06.00—18.00—06.00) the sweat rate was calculated; skinfold thickness was also measured morning and evening, each day. Records were kept of all heat or other illness or injury and military performance was also assessed.

The results are illustrated in Figures 7.6, 7.7, 7.8 and 7.9. There were

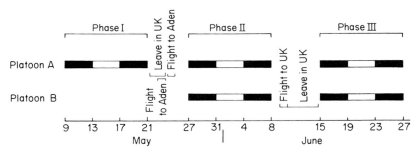

Fig. 7.6 Plan of the experiment. In each phase there was a four-day period of activity followed by four days rest and a final four days of activity. Black blocks active days; White blocks rest days.[7]

significant differences between the acclimatized men from Bahrain and their opposite numbers from the UK. These differences were reflected in the body temperatures during the 'hard' days especially. Each 'hard' day began with a 4-hour march, body temperatures being recorded at the end of each hour. As can be seen from the figures, the body temperatures in the acclimatized group were significantly lower than in the unacclimatized group. The differences between these two groups diminished over the course of the 12 days but there was still a significant difference on the last day. Sweat rates were higher in the acclimatized than in the unacclimatized group and again the differences became less in the course of the exercise. Heart rate of the acclimatized group was the lowest with the difference becoming less during the trial. Physiologically, the acclimatized group had a definite advantage over the unacclimatized group throughout the period in Aden. These differences were much less during the four rest days, when body temperatures were not increased greatly

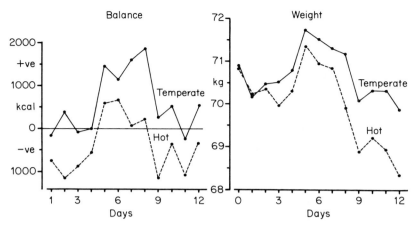

Fig. 7.7 Calorie balance between intake and expenditure and weight changes in temperate and hot climates.

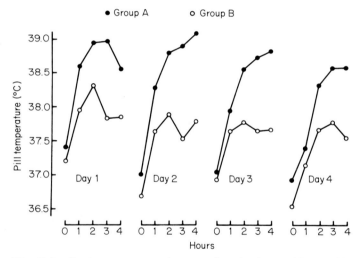

Fig. 7.8 Body temperature in unacclimatized men (Group A) and acclimatized (Group B) during a daily 4-hour march in Aden. On each of the first four days the acclimatized men had lower temperatures than the unacclimatized. There was some evidence of acclimatization developing in both groups.[7]

by activities such as swimming. The most notable difference between the two groups was in the number of heat casualties. Throughout the period of the exercise in Aden, there were only three casualties which could be attributed to heat in the acclimatized group, whereas in the unacclimatized group there were over 30 cases with some men having more than one episode of heat illness or collapse. As far as military performance was concerned, in the first few days it was extremely difficult for the unacclimatized group to complete

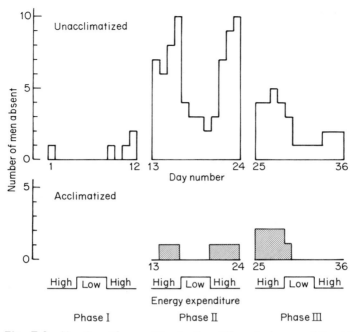

Fig. 7.9 Number of casualties in the different phases of the trial. Phase I was in the UK in which only the unacclimatized participated and there were few casualties. But in Aden in Phase II the unacclimatized had many casualties compared with the acclimatized men.[8]

their 4-hour march and on two occasions they were unable to do so. The acclimatized group, although they did not find the task easy, were able to complete the march without any men falling out. Water intake (as may be expected from the higher sweat rates) was higher in the acclimatized compared to the unacclimatized group although the difference was not great. This difference was, however, significant and persisted during the period in Aden. Food intake was not very different in the two groups, being low in both cases; markedly lower comparing the first phase in the UK to the second in Aden although the energy expenditure estimates, showed that it was virtually identical in Aden to that in the UK. Food intake averaged approximately 4200 J (1000 kcal/day) less in Aden than in the UK. This resulted in loss of body weight in both groups which tended to be rather more in the unacclimatized group.

The third phase was carried out to establish if there was any real difference between these platoons other than their acclimatization to heat. When they were brought back to the UK they carried out exactly the same set of exercises as for the first phase. In virtually all the parameters there was no significant difference between the acclimatized and unacclimatized platoons. It was concluded from this experiment that natural acclimatization to heat confers a significant military advantage which is additional to that due to a high level of physical fitness. It has been recognized for many years that a high degree of fitness improves performance in the heat as in the cold but amongst these men, who were all extremely fit, those who had not been acclimatized to heat were unable to perform as well as those who were.

Since the time of that trial, under the auspices of the International Biological Programme (IBP)[11], studies have been carried out on many people in different parts of the world, including work by Fox *et al.*, using techniques developed at Hampstead[12, 13, 14]. Some results are shown in Fig. 7.10 where it may be seen that there are considerable differences between people living in hot climates compared, not only to those living in cool or temperate areas but also between each other. Sweat rate in New Guinea subjects is much lower than that of the inhabitants of Nigeria. Other differences are also evident from the figure and it appears likely that the acclimatization process is associated with the genetic make-up of the individual. Other IBP studies are summarized in the book edited by Collins and Weiner[11]. A valuable set of references has been compiled by Sciaraffa *et al.*[15]

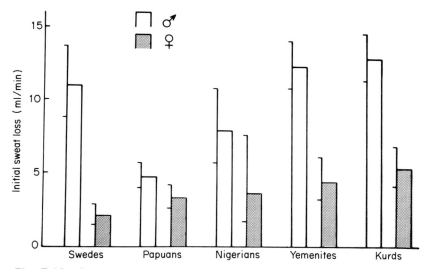

Fig. 7.10 Sweat rates under the standardized conditions of the 'bed' test (see p. 148), measured in people indigenous to various climates. In each group men have higher sweat rates than women. There are significant differences between the groups suggesting the importance of genetic factors.[28]

Acclimatization by raising body temperatures
Fox put forward the hypothesis that the essential stimulus to acclimatization was a rise in body temperature. He then showed that acclimatization could be produced by passive elevation of the body temperature. This was achieved by exposing subjects to very hot conditions, so raising body temperature to between 38.0 and 38.5°C. This temperature was maintained by suppressing the heat loss from the subject, who wore a plastic suit (Fig. 7.11) through which sweat could not evaporate. The rate of sweat loss was controlled by

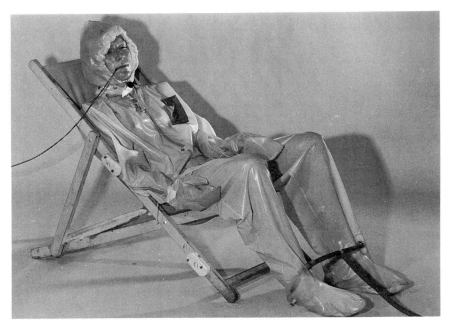

Fig. 7.11 A subject wearing an impermeable plastic suit. Air can flow into the suit through ducts at the ankle. By varying the airflow the evaporation of sweat can be controlled and hence the body temperature which has previously been elevated by exposure to a high temperature.

varying the amount of air flowing into the suit. In this way, Fox was able to maintain a constant body temperature of 38°C with the subject sitting in a chamber kept at 30°C. This technique made it possible to examine the effects of raising body temperature passively to different levels for varying periods of time. The results showed that the degree of acclimatization produced by this method depended on the level to which body temperature was raised, and the length of time for which it was kept at this level on successive days. In a 15-day exposure period, a full degree of acclimatization could be achieved, measured by sweat rate, body temperature and heart rate just as when subjects were exposed in a climatic chamber (Fig. 7.12).

Having shown that the acclimatization process could be stimulated by a rise in body temperature, Fox *et al.* designed a bed[16] by which acclimatization status could be assessed by the measurement of sweat rate when a subject was exposed to heat. In this bed, the subject was enclosed in a plastic suit within which any sweat produced was collected. The outer surface of the plastic suit was covered with a special blanket in which air, at a controlled temperature, was passed through to the surface of the suit. When the subject was surrounded with insulating material, body temperature could be raised to the desired level by adjusting the quantity of hot air passing through the enveloping blanket. The effects of a raised body temperature could be studied within

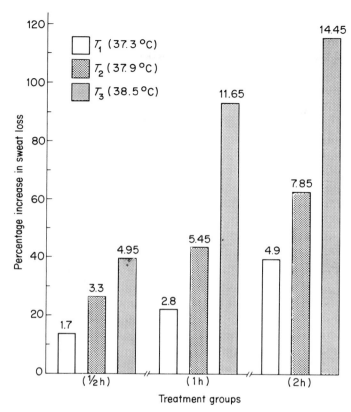

Fig. 7.12 Percentage increase in sweat rate after daily exposure for 15 days to half hour, one hour or two hours elevation of body temperature to 37.3, 37.9 or 38.5°C. The increase in sweat rate depends on the degree of elevation of body temperature and its duration.[28]

this portable bed without the need for expensive and immobile climatic chamber facilities. Sweat rate was assessed by measuring the volume of sweat which collected in the suit; heart and pulse rates were measured at the same time to assess the level of acclimatization. It was by the use of this bed that the results shown in Figure 7.13 were obtained.

Results of Fox's work make it highly probable that his hypothesis is correct: the stimulus to acclimatization is a rise in body temperature, by whatever means this is achieved, such as exercising in the heat.

It was first shown by Nielsen[17] and subsequently confirmed by Lind[18] and others, that the effect of exercise in raising body temperature is independent of the environmental temperature, within fairly wide limits. However, in experiments in climatic chambers, subjects are exposed to temperatures which are beyond these limits and the rise in body temperature is steeper than it would be in more temperate conditions. The conclusion is that whether it is

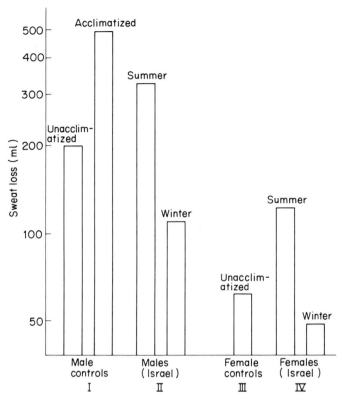

Fig. 7.13 Sweat loss with controlled hyperthermia. I, in male UK subjects before and after acclimatization; II, in male Israeli villagers in summer and winter; III, in female UK unacclimatized subjects; IV, in female Israeli villagers in summer and winter. (From experiments by R. H. Fox *et al.* (1973)[12]).

raised by exercise, by passive exposure in a very hot environment or even in a hot bath, any means by which body temperature is raised will eventually induce the changes which we associate with acclimatization to heat.

Mechanism of sweat rate increase

The next question which Fox tried to answer was, 'What is the mechanism responsible for the most salient feature of acclimatization, that is, the increase in sweat rate?' It could be that the temperature regulating centre becomes more sensitive and there is a stronger stimulus to the sweat glands at a particular level of body temperature. On the other hand, it may be that the increase is due to a training of the sweat glands analogous to the effects of muscle training on performance.

Fox *et al.*[19] examined these possibilities by measuring the sweat rate in the arms of subjects who had their body temperature raised. Following 12 daily periods of hyperthermia the sweat rate in the forearm could increase almost threefold (see Fig. 7.14). In the next experiment the subject had one arm put in very cold water during the time that body temperature was raised, so inhibiting sweating in the cool forearm. Over the period of acclimatization one arm was sweating normally, and the sweat rate was increasing daily; in the other arm sweating was suppressed. At the end of the period of acclimatization the subject was exposed to hot conditions in the climatic chamber with both arms freely exposed to the same environmental conditions. The sweat rate was measured in both arms; in the arm which had been subjected to the effect of a raised body temperature, sweat rate was substantially increased, between two- and three-fold, compared with the initial level before the experiment began. In the experimental arm which had been cooled during the acclimatization period, and which had not produced any sweat, the sweat rate had not changed in comparison with the level before acclimatization (Fig. 7.14). The conclusion from this was that because the temperature regulating centre had been sending out the same stream of impulses to the two arms on the occasion of the final experiment and one arm had responded with a very large increase in sweat rate, and the other showed no increase at all,

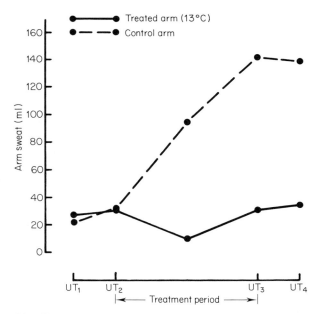

Fig. 7.14 Sweat rates in arms measured in standardized conditions (Uniformity test (UT)$_{1, 2, 3, 4}$). Between UT$_2$ and UT$_3$, the subjects were exposed daily in a hot room with the 'treated' arm immersed in water at 13°C. In UT$_3$ and UT$_4$, the sweat rates in the two arms, both exposed to the hot environment were compared. The control arm had greatly increased its sweat rate compared with the results in UT$_1$ and UT$_2$, whereas the arm kept cold during the treatment period had its sweat rate unchanged. This result indicates that acclimatization to heat is largely due to increased sensitivity of the sweat glands.[19]

the effect could not have been due to changes in the centre but must have been peripheral. Therefore, the overall conclusion from these experiments was that changes in sweat rates are primarily a peripheral phenomenon, that is a training effect analagous to that found in exercise.

Genetic factors

As far as the genetic effects are concerned, most of the work which has been done would indicate that the environment is the most important factor in determining individual sweat rates.

Studies have been carried out in Israel on Yemenite and Kurdish Jews living in adjacent villages, leading similar lives and having similar houses. Detailed observation and measurements of environmental conditions showed that these were virtually identical for the Yemenite and Kurdish Jews[20]. However, there were striking differences between the two populations in their genetic characteristics. The blood group frequencies of the Kurdish Jews were found to differ markedly from the Yemenite Jews[21] as were the plasma-protein systems and the red cell enzymes.

There were also some anthropometric differences between the two communities. The Kurdish Jews were heavier and taller than the Yemenites. Daily energy expenditure and food intake were similar.

The degree of acclimatization to heat in the summer was assessed using the bed designed by Fox. The average sweat rates of the men in the two groups were virtually identical and were consistent with a high degree of acclimatization. The Kurdish and Yemenite women also had similar sweat rates but at a lower level than in the men. In the winter, although the climate was mild with only a few cool days, the sweat rates in both men and women were as low as in non-acclimatized subjects.

The Kurdish and Yemenite Jews, men and women, lived exceptionally similar lives in all respects and were exposed to an identical climate. In spite of their very marked genetic differences their state of acclimatization to heat was remarkably similar[22]. So far, it has not been possible to find an analogous situation elsewhere in the world. The Israel study provides strong evidence that environmental effects are greater than genetic influences as far as acclimatization is concerned. However, even when a group of people who had been exposed to quite similar environmental temperatures were examined, marked individual variations were seen. This is difficult to explain without recourse to genetic factors. One way of investigating this problem has been to examine the sweat rates of twins. This was first done by Fox who examined a number of monozygotic and dizygotic twins, but the results were, to a certain extent, equivocal; it did not appear that twins resembled each other any more closely than brothers or sisters. This indicated that the environment was perhaps more important than the genetic components. More recently Collins and Weiner[23] have been able to examine another group of monozygotic and dizygotic twins and to show clear distinctions between these two groups. Their results showed that the monozygotic twins resembled each other significantly more closely than did dizygotic twins, therefore implying that there was indeed a genetic component. The factor which is extremely difficult to control in such experiments is the precise environmental conditions under which the twins had lived. From Collins's records it would appear that the environmental conditions of the two groups, for each pair, were relatively similar. It would be difficult to explain his results on the effects of environments alone. One may conclude that environmental conditions are, perhaps, the most important in determining sweat rates but that genetic factors undoubtedly contribute.

The extensive studies of animals in hot conditions have been reviewed by Ingram and Mount[24], and by Bligh *et al.*[25]. There are also many references to animal work in Hensel[26].

The behavioural responses of animals are not unlike those observed in man. Hardy[27] considered that animals in general relied on behavioural thermoregulation and only utilized autonomic mechanisms if conditions prevented or made it difficult to utilize behavioural patterns.

References

1.   Scholander, P. F., Walters, V., Hock, R. and Irving, L. (1950). Body insulation of some arctic and tropical mammals and birds. *Biological Bulletin* **99,** 225–36.
2.   Scholander, P. F., Hock, R., Walters, V. and Irving, L. (1950). Adaptation to cold in arctic and tropical animals and birds. *Biological Bulletin* **99,** 259–71.
3.   Wolff, H. S. (1958). A knitted wire fabric for measuring mean skin temperature and for body heating. *Journal of Physiology* **142,** 1–2P.
4.   Macpherson, R. K. (1960). *Physiological responses to hot environments*. Medical Research Council Special Report, No. 298. HMSO, London.
5.   Hellon, R. F., Jones, R. M., Macpherson, R. K. and Weiner, J. S. (1956). Natural and artificial acclimatization to heat. *Journal of Physiology* **132,** 559–76.
6.   Ladell, W. S. S. (1964). Terrestrial animals in humid heat. In: *Handbook of Physiology*, pp. 625–59. Ed by D. B. Dill. American Physiological Society, Washington.
7.   Edholm, O. G., Fox, R. H., Goldsmith, R., Hampton, I. F. G. and Pillai, K. V. (1965). A comparison of heat acclimatisation in Indians and Europeans. *Journal of Physiology* **177,** 15–16P.
8.   Edholm, O. G., Adam, J. M., Cannon, P., Fox, R. H., Goldsmith, R., Shepherd, R. D. and Underwood, C. R. (1961). *Acclimatisation to heat; a study of artificially acclimatised, naturally acclimatised and non-acclimatised troops on exposure to heat.* Report No. APRC 61/25. Army Personnel Research Committee, Medical Research Council, London.
9.   Edholm, O. G., Fox, R. H., Adam, J. M. and Goldsmith, R. (1963). Comparison of artificial and natural acclimatization. *Federation Proceedings* **22,** 709–15.
10.   Edholm, O. G., Fox, R. H., Goldsmith, R., Hampton, I. F. G., Underwood, C. R., Ward, E. J., Wolff, H. S., Adam, J. M. and Allan, J. R. (1964). *The effect of heat on food and water intake of acclimatised and non-acclimatised men.* Report No. APRC 64/16. Army Personnel Research Committee, Medical Research Council, London.
11.   Collins, K. J. and Weiner, J. S. (1977). *Human adaptability.* Taylor and Francis Limited, London.
12.   Fox, R. H., Even-Paz, Z., Woodward, P. M. and Jack, J. W. (1973). A study of temperature regulation in Yemenite and Kurdish Jews in Israel. *Philosophical Transactions of the Royal Society of London Series B* **266,** 149–68.
13.   Fox, R. H., Budd, G. M., Woodward, P. M., Hackett, A. J. and Hendrie, L. (1974). A study of temperature regulation in New Guinea people. *Philosophical Transactions of the Royal Society of London Series B* **268,** 375–91.

14.   Budd, G. M., Fox, R. H., Hendrie, A. C. and Hicks, K. E. (1974). A field survey of thermal stress in New Guinea villages. *Philosophical Transactions of the Royal Society of London Series B* **268**, 393–400.
15.   Sciaraffa, D., Fox, S. C., Stockman, R. and Greenleaf, J. E. (1980). *Human acclimation and acclimatization to heat: a compendium of research 1968-78.* NASA Technical Memorandum 81181.
16.   Fox, R. H., Crockford, G. W., Hampton, I. F. G. and MacGibbon, R. (1967). A thermoregulatory function test using controlled hyperthermia. *Journal of Applied Physiology* **23**, 267–75.
17.   Nielsen, M. (1938). Die Regulation des Korpertemperatur bei Muskelarbeit. *Skand. Arch. Physiol.* **79**, 193–230.
18.   Lind, A. R. (1963). A physiological criterion for setting thermal environmental limits for everyday work. *Journal of Applied Physiology* **18**, 51–6.
19.   Fox, R. H., Goldsmith, R., Hampton, I. F. G. and Lewis, H. E. (1964). The nature of the increasing sweating capacity produced by heat acclimatization. *Journal of Physiology* **171**, 368–76.
20.   Edholm, O. G. and Samueloff, S. (1973). Biological studies of Yemenite and Kurdish Jews in Israel. *Philosophical Transactions of the Royal Society of London Series B* **266**, 85–95.
21.   Tills, D., Warlow, A., Mourant, A. E., Kopec, A. C., Edholm, O. G. and Garrard, G. (1977). The blood groups and other hereditary blood factors of Yemenite and Kurdish Jews. *Annals of Human Biology* **4**, 259–74.
22.   Edholm, O. G., Samueloff, S., Mourant, A. E., Fox, R. H., Lourie, J., Lehmann, H., Lehmann, E. E., Bavly, S., Beavan, G. and Evan-Paz, Z. (1973). Biological studies of Yemenite and Kurdish Jews in Israel. Conclusions and Summary. *Philosophical Transactions of the Royal Society of London Series B* **266**, 221–4.
23.   Collins, K. J. and Weiner, J. S. (1979). Thermoregulation in twins. *Annals of Human Biology* **6**, 290–1.
24.   Ingram, D. L. and Mount, L. E. (1975). *Man and Animals in Hot Environments.* Springer-Verlag, Berlin and New York.
25.   Bligh, J., Cloudsley-Thompson, J. L. and Macdonald, A. G. (Eds) (1976). *Environmental Physiology of Animals.* Blackwell, Oxford.
26.   Hensel, H. (1981). *Thermoreception and Temperature Regulation.* Monograph of the Physiological Society. Academic Press, London and New York.
27.   Hardy, J. D. (1972). Peripheral inputs to the central regulator for body temperature. In: *Advances in Climatic Physiology.* Ed by S. Itoh, K. Ogata and H. Yoshimura. Igaku Shoin, Tokyo.
28.   *Principles and Practice of Human Physiology* (1981). Ed. O. G. Edholm and J. S. Weiner (1981). Academic Press, New York.

8

Responses to Cold

Introduction

One of the main ways in which the body tries to combat the effects of cold is to increase heat production. This is achieved most evidently by shivering, but apart from this, muscle tensing together with an enhanced metabolism will increase heat production.

A number of experiments have been carried out, notably by Yoshimura[1] in Japan and by Wyndham in South Africa[2], with subjects resting naked in climatic chambers at environmental temperatures ranging from 30°C down to as low as 10°C. Figure 8.1 illustrates the results obtained in a climatic chamber with little air movement; a small increase in heat production occurred when the environmental temperature was lowered from 30°C to 25°C.

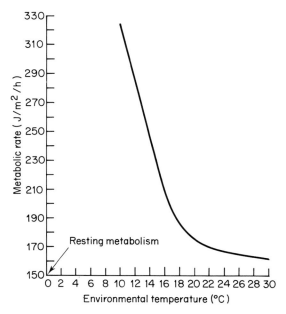

Fig. 8.1 Variation of metabolic rate with environmental temperature. Average values from Yoshimura[1] and Wyndham[2].

Below 25°C, a further small increase in heat production was observed but this was not really of any consequence until the environmental temperature reached 20°C. At this point, the heat production increased abruptly and continued as the temperature was lowered to 15°C or even 10°C. In these studies, it was necessary to follow the metabolic rate for some time after changing the environmental temperature as the effects, at least at intermediate chamber temperatures, were not immediate. One of Yoshimura's experiments showed that the effect may take 30 minutes or more to become fully developed. Heat production increases slowly at environmental temperatures of 20–21°C and will not achieve a steady state for some time. The curve shows that a nude man at rest has to increase his heat production if body temperature is to remain constant in an environment cooler than 25°C. Even when conditions are warmer than this the body temperature is likely to fall, although in the short term there may be small metabolic compensations.

There is an increased heat production in shivering muscle and the muscles may also become tense causing the subject to feel stiff, especially in the cold. An important phenomenon known as 'non-shivering thermogenesis' is a collection of mechanisms which have been clearly demonstrated in animals that have been acclimatized to cold and who show an increased heat production with a very minor degree of shivering.

The heat generation occurs mainly in the liver but is also found in a special tissue described as 'brown fat'. This term is purely descriptive in that the fatty tissue has a brown colour. It is found in a variety of newborn mammals and has been widely studied in the rabbit. Brown fat is found at the back of the neck and between the shoulder blades but can also be identified in other parts of the body. It has a very rich blood and nerve supply and the metabolism in these fat cells can, in effect, be turned on by stimulating the nerve supply; it may also be activated by the action of circulating catecholamines. The animal possessing brown fat has a reserve of heat which can be produced very rapidly, and when this heat is generated at the back of the neck it can effectively raise the temperature of the blood supply to the brain.

In man, there is evidence of brown fat in the infant but it seems extremely unlikely, at least with our present knowledge, that it persists into adulthood, or that it can be regenerated in the adult who is exposed to cold. In the human infant it may be a very valuable source of heat but it is suggested that most of the brown fat disappears within the first year of life although small quantities may persist in the intra-scapular region and also in the perirenal fat. It has also been suggested, but on very speculative evidence, that there might be an increase in the amount of brown fat in the elderly compared with younger people and that this could be one means by which an old person, whose shivering mechanism tends to diminish, could maintain heat production[3].

The hedgehog has a very good store of brown fat and it is interesting that the mobilization of this fat may enable him to come round to consciousness after hibernation. An elderly hedgehog may not regain consciousness after hibernation because of an inadequate supply of brown fat which has been diminishing consistently with advancing age[4].

Shivering

The phenomenon of shivering is produced by the changes that occur when the skin is cool. Shivering is generally initiated by a lowering of skin temperature without there necessarily being a fall in body temperature. However, if, in addition to skin cooling, the body temperature does fall then the shivering can become much more violent. Shivering increases to a maximum which occurs at a body temperature of around 34–35°C. At temperatures below this, shivering usually becomes less marked and by the time 32–33°C is reached it will almost certainly have completely stopped. The changes that occur in shivering are preceded by what has been described as thermal muscular tone.

Measurements have now been made in animals and in man to show that the changes in the firing rates of motor units increase with cooling but without any of the actual movements that one associates with shivering; there is no tremor or actual motion, just an increase in muscle activity. This is the basis for the feeling of stiffness that most of us experience when we get cold, making it harder to move the limbs and to make muscular movements. With a further stimulus, either by cooling the skin or by the deep body temperature beginning to fall, frank shivering begins. At this stage, the muscle units fire at different frequencies of repetition and out of phase with each other. With further cooling, some of these discharges become more synchronized to produce groups of fibres whose muscle units discharge at the frequency of a tremor which is usually between 10 and 12 per second. This eventual co-ordination of the various motor units is due largely to the periodic inhibition of the impulses to the motor nerves which are almost certainly due to the regular discharges in the fibres of the muscle. If the dorsal roots in a cat are cut then the tremor of shivering is abolished but not the thermal muscular tone.

During shivering the changes that occur in overall heat production can be quite considerable and show an increase of the order of five to six times the resting metabolic rate. However, this increase is seldom maintained for more than a few minutes and even the most violent shivering is characterized by periodic bursts of shaking interspersed with periods of almost complete rest. The heat production therefore varies rapidly from the very high level down to levels which may be little higher than the resting value itself. When averaged over a period of time, even with considerable body cooling, shivering probably does not increase the body heat production by more than about 50–100 per cent and this is about the level that can be maintained over a period of an hour or more.

From a thermodynamic point of view shivering is rather more efficient than voluntary muscular contraction in terms of heat production. Something of the order of 60–70 per cent of the energy used in a voluntary contraction of muscle appears as heat; the rest is not immediately available unless by internal friction in the muscles or in the environment but in shivering there is an immediate transformation of all the energy involved to heat as the limbs do not effectively move. Since both agonist and antagonist muscles increase in tension no external work is done. It is of course true that the heat produced is directed towards the surface of the body, and therefore heat loss by conduction and via blood flow to the skin is rather high. Even so, shivering is a very effective means of maintaining body temperature. It may be worth pointing out that shivering can occur in muscles which are paralysed through a loss of the normal motor control such as can happen in a cerebral haemorrhage when the pyramidal tract is damaged[5].

The pathway is not through the pyramidal tract but is either by the lateral tecto-spinal or rubro-spinal tract. It is perhaps for this reason that shivering in a particular muscle is generally suppressed during its voluntary activity. Although the overall heat production is not very high it is effective and there is one aspect in which it is of some importance. This is when the muscles of the body do not all shiver at once. Slonim[4] has shown in a series of studies on animals which were cooled, that the first muscles to shiver were in the neck and this also appears to be true in man. The effect of this is to produce heat to warm the blood flow to the brain as the neck muscles are adjacent to the blood supplied to the brain via the vertebral arteries coming up from the spinal cord. The relevant arteries pass straight through to the brain and join the circle of Willis, supplying something of the order of one-third of the blood supply to the brain. If the neck muscles shiver they will raise the temperature of this blood and Slonim has shown that this is a feature that can be found in a number of animal species.

Shivering may be affected by factors other than temperature; it can be depressed by anoxia or rather by a lowering of the partial pressure of oxygen. Similarly, it may be inhibited by carbon dioxide. In a well stirred water bath shivering usually begins at a water temperature below 33°C. If the water is saturated with carbon dioxide and thus made into soda water, shivering is inhibited both by the decreased sensation of cold and also by the effect of the carbon dioxide on the conductivity of the water.

Anti-pyretic drugs suppress shivering and substances containing magnesium ions, for instance, also inhibit the effect. This is not due to the curare-like effects of magnesium since it occurs at a very low concentration. It is probable that it has a specific effect as it is also related to the onset of hibernation and changes in magnesium level have been measured before and during hibernation. Shivering is also inhibited by the injection of insulin presumably due to a fall in blood sugar level.

Not only is shivering sensitive to the level of oxygen and carbon dioxide but it is also inhibited by any form of anaesthesia however mild. During anaesthesia there is a great susceptibility to falling body temperature. If deep temperature is measured during surgical operations it is not uncommon to observe quite marked changes unless the operating theatre is very warm (see p. 230).

Hypothermia

Exposure to low air temperatures, combined with increasing air movement can lead, in the absence of adequate clothing, to such a high heat loss that body temperature falls. A state of hypothermia is said to exist when body temperature is 35°C or below. Accidental hypothermia can be due to exposure in bad weather or to immersion in cold water or, particularly in the elderly, it may be a consequence of illness or accident; in addition, hypothermia may be induced for surgical reasons.

It is more difficult to lower body temperature than to raise it and when this is done deliberately rigorous cooling techniques are required. There are a variety of methods used to induce hypothermia prior to operation, ranging from packing ice around the body to extra-corporeal circulatory heat exchange. The patient is anaesthetized during body cooling not only to allay pain and distress but also to prevent shivering and thereby achieve body cooling without compensatory heat production. It is the efficacy of shivering that normally makes hypothermia relatively rare.

Stages of hypothermia
Modest falls of body temperature to 36°C are common and occur in many individuals during sleep[6]. Small falls do not necessarily stimulate metabolic responses as they are often associated with quite high skin temperatures. When asleep in bed with adequate covering there is peripheral vaso-dilation and minimal muscle activity. Heat production falls and body temperature drops. There is little evidence of physiological response to such falls although there is some slowing of the heart and respiration. Greater falls of deep body temperature to 35°C or below are usually associated with more evident physiological responses. However, it is not possible to dissociate the environmental conditions and bodily changes.

As mentioned above, hypothermia can be due to accident or illness and the varying antecedent conditions need to be considered separately. The general effects of various degrees of body cooling can be described, although there is very considerable individual variation. At a body temperature of 35°C there is usually violent shivering which occurs in bouts and the increased heat production may be sufficient to maintain body temperature depending on the rate of cooling; but high levels of shivering cannot be maintained indefinitely so that body temperature will begin to fall. As it falls below 35°C most subjects will have difficulty in controlling their movements. This is because the temperature of the muscles will also become lower, movements will become increasingly impaired and the subject may fall frequently or become clumsy. These effects are due directly to changes in muscle temperature.

As far as the individual himself is concerned, apart from clumsiness there will also be a beginning of clouding of consciousness at a body temperature of 34°C. At 33°C consciousness is most definitely impaired and it would seem that there is a loss of sensory, particularly visual, information. Sight may become blurred and the subject responds with difficulty to stimuli. At 31–32°C there is clouding of consciousness in all subjects and in some cases consciousness may be lost altogether. At temperatures of 30–31°C almost all subjects will be unconscious.

Hypothermia at this level and below increases the risk of mortality from ventricular fibrillation of the heart. The cause of such fibrillation is far from certain but it appears to be largely due to the direct effect of temperature on the pace-maker. There is also an effect due to the movement of potassium from the cells into the extracellular fluid. When body temperature drops below 31°C death may occur at any time due to ventricular fibrillation.

However, temperatures much lower than this have been achieved in surgical conditions when they have been maintained for operations such as repair of the heart or operations on the brain. At temperatures below 20°C which have been maintained for periods of hours, the subjects will have a very slow or virtual cessation of heart beat. The brain can survive without blood flow for a very much longer time at such low temperatures than at normal temperatures. The reason that these low temperatures can be achieved is because the heart's ventricular beat can be restored using a defibrillator. This can be done directly to the heart or through the chest wall, and not only restores the normal heart beat but will keep it beating as body temperatures cool to levels of 20°C or lower. At such temperatures, heart rate has been decreasing more or less in step with the fall in body temperature. The respiration rate also falls and there is some fall in blood pressure. There is an increase in the peripheral resistance due to the greater viscosity of blood as the temperature falls. At about 15°C cardiac action usually stops altogether and at this stage, in one sense, the patient may be described as dead because no electro-encephalogram activity can be recorded. However, such a patient can be restored to life by rewarming and this has been done successfully even from temperatures as low as 10°C provided that they are not maintained for too long.

Hypothermia causes some loss of consciousness before there is the danger of ventricular fibrillation. An experiment which was carried out to see how well a low body temperature could be maintained over a period of time led to some unusual and unexpected results. The purpose of producing hypothermia was to determine whether it was possible to treat some cases of cerebral tumour. Astrocytomata are a very lethal form of tumour and do not respond to deep X-ray treatment, but it was thought that at low body temperature the tumour cells might become more sensitive than the rest of the brain cells and therefore be more susceptible to the X-rays. A number of patients were kept at a body temperature of between 30 and 31°C for up to three weeks. At these temperatures patients recovered consciousness and were able to converse, eat, drink, read the newspaper and indeed one patient was sufficiently well to try a flirtation with a nurse! When the patients were subsequently rewarmed they had no memory of what had happened to them at the low body temperature. They could not remember what they had been reading or any of their conversations; all that remained was a blurred recollection that something had happened during that period.

Accidental hypothermia

One of the causes of accidental hypothermia[7,8] in young people in the UK is exposure when climbing in the hills of Wales or Scotland or walking over the moors in the north of England. Many young people have been exposed to very severe climatic conditions and it is not often realized that the climate in the UK in winter can be lethal. In such cases of accidental hypothermia the victims will go through some of the stages described above. In wet and windy conditions, when wearing inadequate clothing, body temperatures will begin to fall. Once this begins, the ability to walk or climb is impaired because of increasing muscle weakness. As this weakness involves a diminution of muscle activity, heat production begins to fall, and this accentuates the fall in body temperature. In this way, the climber or hill walker who is inadequately clothed will be at great risk when meeting severe weather conditions. Nevertheless, even though he may be quite close to death, that is with a body temperature of around 30–31°C or lower, he may also be restored remarkably quickly. In one episode a young man was brought to a farmhouse unconscious and apparently dead and was put directly into a bath of hot water at over 40°C. He recovered rapidly and within about an hour was conscious and able to move about without any difficulty.

Accidental hypothermia of hill walkers and climbers can be reversed provided that action is taken quickly. Most of the fatalities that are described as being due to exposure are actually due to hypothermia. The incidence of such accidents has decreased in recent years as it has become realized that proper clothing must be worn during long periods out of doors, not only in winter months but also in the spring and autumn[9].

There is considerable individual variation in the sequence of events during body cooling but impairment to the awareness of dangers and difficulties occurs at about 35°C and lower. It is these levels of temperature which often account for the tragedies of hill climbers and walkers who may be lost or delayed in their progress and who become exposed to severe conditions which result in body cooling. As this occurs, movements become much more clumsy and people are apt to stumble; in hill walking, falls can ensue which may themselves be serious; the individual himself is very often unaware of the ways in which his own mental processes have been affected. If he is spoken to, roused and understands, he may be quite indignant at being told that he must now take shelter or some action to prevent further loss of heat. The lack of awareness of deteriorating skill is reminiscent of the effects of hypoxia on performance, where people may be sure that they are doing perfectly well even though they may be suffering from obvious hallucinations. Amongst those who have survived such experiences with exposure to cold it is not infrequent to have reports of blurred vision and visual delusions.

An interesting observation in connection with this is that the temperature of saturated clothing in the presence of moderate air movement is that which a wet bulb thermometer would read. This in turn leads to a very high rate of evaporative heat loss to the environment accentuating the fall of deep body temperature[10].

In an experiment carried out in the climatic chambers at Hampstead, subjects wore adequate clothing to maintain body temperature whilst sitting at rest at 5°C with an air movement of 2 m/s. They then worked on a bicycle ergometer and increased their heat production. Body temperature rose slightly but stabilized and did not continue to rise. These subjects had their clothing saturated with water by standing under a shower and they then felt extremely cold. After riding on the bicycle ergometer and increasing heat production they were unable to maintain body temperature in spite of an oxygen consumption which increased to the order of 2 litres/min. Body temperature fell steadily throughout the exercise period and it could be calculated that the insulation of the clothing, saturated as it was with water, was virtually nil and they were little better off than being completely nude.

Hypothermia in the elderly

Elderly subjects are more liable to accidental hypothermia than younger age groups because with ageing, in many people although not all, there is an increasing impairment of temperature regulation. This means that elderly subjects do not necessarily shiver if they are sitting in a cold room. As a result of this an elderly person may not complain of or feel the cold but will be losing heat at such a rate that body temperature will fall. Many surveys have been carried out to study the effects of particular environments on the elderly, and examinations of old people living in their own homes during the winter revealed a number of cases with body temperatures between 35 and 36°C, but where the subjects themselves did not complain of feeling cold. A full description has been given by Fox and his colleagues of such a survey[11].

One of the problems with the elderly is that the peripheral circulation is not as well controlled as in younger people and Collins *et al.*[12] have shown that in some cases skin blood flow did not decrease in the cold, that is, there was no effective vaso-constriction and this helped to account for the dangers to the elderly of cold surroundings. Another age group which is also vulnerable to cold is the very young, particularly young babies under six months old. Babies do not shiver; at birth they have a moderately good temperature regulating system. Their blood vessels will constrict in the cold and dilate in the heat and they may even sweat a little but are unable to shiver. The infant is susceptible to hypothermia if unduly exposed to cold. One of the dangers of the cold baby, and the reason why the condition was only fairly recently recognized, is that in the cold the baby will look as if it has a fine pink or red complexion. The blood vessels in the skin are dilated but the blood itself, because it is so cold, is not reduced and instead of looking blue the cold baby looks pink or red and this has unfortunately led to a number of fatalities.

The treatment of hypothermia is outlined by Golden[13] and two views are put forward by Emslie-Smith *et al.*[14] and by Ledingham[15]. Collins[16] considers the whole field of hypothermia, including treatment.

As Emslie-Smith *et al.* state 'There cannot be any standard'—in the management of hypothermia, particularly as it is common to find patients also suffering from other frequently serious conditions.

In the field situation the victims of hypothermia are generally young and physically fit. Rapid rewarming is still the treatment of choice. In hospital, cases are in various categories. Mild hypothermia, uncomplicated by other illness, can be treated with gentle warmth, allowing body temperature to increase metabolically.

When the patient is elderly and obviously ill, treatment may be conservative or aggressive, the latter including active rewarming, administration of oxygen and intravenous saline. Conservative treatment is mainly directed against the concurrent illness. There is as yet no general agreement about the most effective regime[12].

Heat transfer in water

The physical principles of heat loss from the body surface in air and water are in many respects similar. Heat is conveyed to the fluid adjacent to the skin by direct conduction and this in turn leads to the formation of a boundary layer which carries the heat away from the surface by convection. The difference between convection in air and water is solely due to the physical properties of the two fluids and it is of interest to compare the convective flows by considering the specific example of a cylinder either in air or water at a temperature of 10°C with the cylinder surface being maintained at 15°C.

Although the thermal conductivities of water and air are quite different they are not sufficient to explain the 200-fold increase of heat loss that occurs when a body is immersed in water. The ratio of thermal conductivity of water to air is about 24 and in order to account for the very large increase in overall heat loss in water it is necessary to understand the mechanism of the formation of the boundary layer flow and the relationship between it and heat transfer.

The natural convection boundary layer around the human has been discussed in Chapter 4 where it was seen that increased buoyancy of the fluid adjacent to the surface produced a convective envelope around the body. In water, a similar effect occurs and the extent and speed of the flow is a function of the physical properties of the fluid. In particular the parameters that determine the flow patterns are the thermal conductivity together with the Prandtl (Pr) number which is non-dimensional and depends on temperature; it is defined as the ratio of the kinematic viscosity to the thermal diffusivity of the fluid (μ Cp/k where μ is the viscosity, k the thermal conductivity and Cp the specific heat of the fluid). In general, at moderate temperatures, the Prandtl number for air has the value of 0.72 and for water the value 10.

The non-dimensional temperature and velocity profiles for free convection have been calculated by Ostrach[17] for a range of Prandtl numbers and Fig. 8.2 shows profiles at 0.3 and 1.5 m height on a cylinder in air and water. The boundary layer in water is similar in extent and in velocity at the two heights with a maximum speed of between 0.12 and 0.13 m/s occurring approximately 0.2 cm from the cylinder surface. However, the profiles for the cylinder in air are quite different at the two levels. At 30 cm the velocity of the air is just over 0.12 m/s and the profile is considerably thicker than for water, but at a height of 1.5 m the boundary layer in air is very much faster than in water at just over 0.28 m/s. This height is almost the limit for the boundary layer flow in air remaining laminar; at greater heights the flow breaks up to become turbulent with changes in its characteristics.

Further examination of the non-dimensional profiles determined by Ostrach enable the local and overall convective heat transfer coefficients to be evaluated. The Nusselt number is often used; this is a non-dimensional heat transfer coefficient and equals $h_c L/k$ where h_c is the coefficient of heat exchange (W/m². °C), L is a characteristic dimension of size, and k is again the thermal conductivity of the fluid. Evaluation of Nusselt numbers leads to an average heat transfer coefficient for the cylinder in air and water. When these values are calculated for a cylinder temperature of 15°C and air or water at 10°C the overall heat transfer coefficients are 1.67 W/m². °C for air and 343 W/m². °C for water. There is thus a factor of 205 between the heat lost by convection in water compared to air.

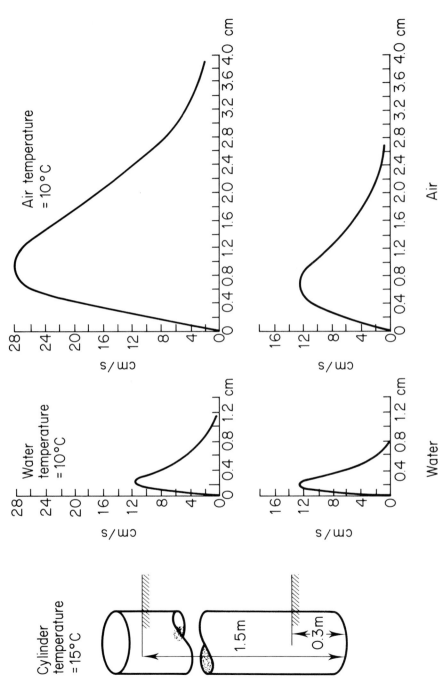

Fig. 8.2 Convective boundary layer velocity profiles in air and water at heights of 0.3 m and 1.5 m on a vertical heated cylinder.

Effects of immersion in water

When long-distance swimmers are studied it is found that they can maintain their body temperature even in relatively cold water. The way in which this is achieved is by a high rate of heat production during swimming. If, instead of swimming, the subject stays still in cold water body temperature may still be maintained and in this case there is a marked difference between individuals due to differences in subcutaneous fat thickness. Fat has a low thermal conductivity compared with muscle or skin and hence a layer of subcutaneous fat can act as an efficient insulator.

One study carried out by Pugh and Edholm[18] was to examine the effects of immersion in water on body temperature. After the Second World War it was shown that survival from ship-wreck depended to a large extent on the water temperature in which the survivors were immersed. At temperatures below 5°C survival time did not exceed 1 hour; at temperatures of 15°C this might be as much as 5 hours but seldom more. With these findings in mind long-distance swimmers were investigated; at that time there was an annual race across the English Channel and the time taken, by the best swimmers, was not less than 12 hours. Water temperature in the Channel, even in summer, seldom exceeds 15–16°C except very close to the shore where water may be washing over relatively warm sand. However, in spite of these low water temperatures the swimmers were apparently able to maintain a sufficiently high body temperature to achieve a crossing. As soon as they were examined it was evident that they had one factor in common; they were all fat (see Figure 8.3). When skinfold thickness was measured this was confirmed and the average subcutaneous fat thickness was considerably higher than the

Fig. 8.3 A group of long-distance swimmers who took part in a race across the English Channel in 1951. It can be seen that all of the swimmers have a thick layer of subcutaneous fat.

average of the population for the same age and sex. This provided part of the answer; the other part of the equation was the rate of heat production and these swimmers, being fine athletes, were able to maintain a high rate of swimming (of the order of 2–3 knots) for upwards of 12 hours. This requires a high rate of metabolic heat production and it was calculated that crossing the Channel may cost in the order of 60 000 J. When detailed measurements were made, partly during the race and partly at the end, it was found that in many cases of swimmers who were forced to retire after varying periods in the water, body temperatures were at hypothermic levels of 35–35.5°C. Some who gave up after considerable periods in the water were, on occasions, confused and would swim in circles, refuse to be taken out of the water, would complain of strange monsters and suffered other obvious hallucinations. Such subjects again appeared to have significant falls of body temperature.

In view of the circumstances it was often difficult to obtain sufficiently detailed or accurate measurements of body temperature and so experiments were made under more controlled conditions on a number of subjects with varying thicknesses of subcutaneous fat. Included in these subjects was one of the Channel swimmers (JZ) who had a very thick layer of body fat as may be seen in Figure 8.4(a) and (b). Experiments were carried out comparing his responses when lying in a bath of cold water with that of another subject (GP) who had a very thin layer of subcutaneous fat. It was found that the fat subject was able to lie in a cold bath at a temperature between 12 and 15°C with apparently no discomfort. A very mild shivering was evident which caused a slight rippling of the water but he was content to lie in the water and read a newspaper. At the end of 2 hours there was no change in his body temperature.

It was quite different with GP. He started to shiver violently soon after getting into the water and his metabolic rate increased approximately five-fold. However, his body temperature began to fall soon after immersion in the water and continued to fall until it reached 34–34.5°C, when it was decided he must be removed from the bath. By that time his shivering had become more intermittent and his overall heat production had diminished. This striking difference between the subjects could not be attributed to differences in physical fitness; both subjects were, by ordinary standards, very fit people and well able to sustain exercise for a long time. The main difference was in their insulation (Figure 8.4(b)). This finding, highlighting the importance of the insulation of subcutaneous fat which provides the reason for individual differences, has since been confirmed by a number of observers[19].

Hayward and Keatinge[20] determined the water temperature in which their subjects could stabilize body temperature. The water temperature varied from 32 to 12°C. The two factors which had the greatest influence on the stabilization temperature were subcutaneous fat thickness and metabolic response.

As a result of these and other experiments it was concluded that the advice which should be given to those who, for one reason or another, are immersed in the sea, is that unless the coast or other refuge is within a close distance (5–10 minutes swimming time away) it is best for the survivor, instead of swimming or making any movements, to float and be as still as possible, when heat

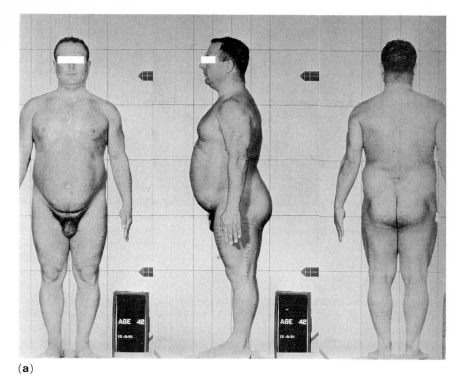

(a)

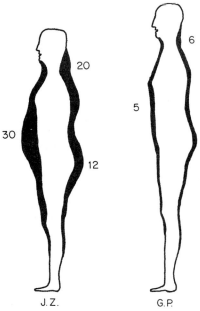

J. Z.

Weight 97 kg.

Height 1.7 m

G.P.

75 kg.

1.8 m

(b)

Fig. 8.4 (a) One of the cross-channel swimmers (JZ) who had a very thick layer of subcutaneous fat all over his body and who had a remarkable tolerance of cold water. (b) silhouettes of JZ and GP with numerals indicating the thickness of subcutaneous fat (mm) at the corresponding points. The height of the figures is to scale but not the thickness of fat. (Reproduced with permission from Pugh and Edholm (1955)[18]).

loss will be diminished and survival time increased. Golden[13] has pointed out that in considering problems of immersion in sea or fresh water, the most immediate danger is that of drowning. However, it is important to realize that in most cases of apparent drowning body temperature is found to be seriously affected. If all the exertions are concentrated on maintaining breathing it can be that the subject may die of hypothermia rather than the immediate effects of drowning.

There is another well-known effect of immersion in cold water which still is not fully explained. During the Second World War when there were a number of victims who were rescued from immersion in cold water it was found that immediately after removal from the sea some became unconscious and died. In the experiments which Behnke[21] carried out on himself with one volunteer, it was shown that after removal from immersion in cold water body temperature would show a drop before recovering during the rewarming period. This 'after-drop' of temperature has therefore been regarded as the cause of fatality after rescue. The suggestion of the 'after-drop' being responsible for fatalities has been examined in detail by Golden and Hervey[22]; they carried out an ingenious experiment using pigs to show that the after-drop of temperature cannot be explained in the way that Burton and Edholm[23] had done. That is, that on coming out of cold water there is an increased blood flow through the skin which is extremely cold and produces a quantity of very cold blood returning to the heart acting directly on the pace-maker and on the ventricular muscle, possibly causing ventricular fibrillation.

Golden has shown that this is a very unlikely explanation; after removing pigs from cold water there was no fall in the temperature of the blood returning to the heart although there was an after-drop of temperature as measured in the rectum. In further experiments, in which he stopped the heart at the moment of removal from the cold water there was also an after-drop of temperature in spite of the rewarming procedure and there was no difference in the temperature changes in pigs with undamaged hearts compared with those in which the heart had been stopped. It therefore seems highly unlikely that this after-drop of temperature can be explained by changes in the circulation and Golden has proposed that this effect should be explicable in terms of physics rather than physiology. A contrary view has been expressed by Collins et al.[24].

The findings on swimmers is similar to the experience of hill walkers and climbers who become lost in severe weather conditions. The way in which the balance between heat production and heat loss is maintained is crucial. In the cross-channel swimmers with their very high rate of heat production, body temperature could be maintained but the results suggest that as soon as the swimmers became tired and their swimming rate decreased (i.e. their heat production was reduced) then body temperature began to fall. As soon as this fall began, muscle temperature also started to fall and the efficiency of muscular movement became impaired; hence the rate of swimming fell even further. Rapidly the situation for the swimmer became physiologically intolerable. Heat production would fall as well as body temperature which would lead to a further fall in heat production and therefore, there could be a very rapid change in overall heat balance. The same finding appeared to be true from the experiences of hill walkers, when it could be shown that individuals might be sufficiently conscious to be able to stumble along and even talk, and yet perhaps half an hour later they would be unconscious and die shortly after unless some means of rewarming was at hand in order to restore body temperature. The balance between heat production and heat loss again appeared to be critical.

Temperature regulation in deep diving

Water temperature is an important factor in the environment of the diver. The thermal effects of immersion in water have already been described and are due to the greatly increased rate of heat loss. Using compressed air, divers cannot safely exceed depths greater than 100 m. A variety of gas mixtures have been used to extend the divers' capacity to descend to greater depths; at present the most commonly used is a mixture of oxygen and helium. A detailed discussion of the physiological problems of diving and the physical laws affecting respiration can be found in Elliott[25].

Helium has a high heat capacity and there is an increase in respiratory heat loss and in addition, since the diver inside his impermeable suit is surrounded by oxy-helium, there is a high rate of convective heat loss. Consequently it is common practice to heat the gas supply to the diver, and if he is wearing a wet suit to heat the water circulating in the suit. The thermal situation is a strange one and involves risks of either hypo- or hyperthermia. The environmental thermal comfort zone becomes very narrow; 32°C can be the threshold for shivering and 34°C for sweating. This narrow band of 2°C may be attributed to the high rate of convective heat loss which is discussed in the next section but more studies are needed on the thermal balance and of the sensations of heat or cold. These are features of thermal comfort which, in ordinary circumstances, are difficult to explain. It may be that our lack of comprehension is due to our continued reliance on the measurement of skin temperature rather than heat flow.

Burton and Edholm[23] described observations concerning heat transfer rates in air which may be relevant to the diving problems. These studies were on the effects of increasing the insulation of clothing on subjects trying to sleep in a climatic chamber kept at −20°C. With moderate insulation, the subject shivered and felt cold but there was no fall of body temperature. With increased insulation the subjects felt comfortable and went to sleep only to wake up after some 3 hours feeling miserably cold and shivering violently. Body temperature at this stage had fallen significantly. Burton argued that with the thick insulation the gradient of temperature from the skin to the outer clothing surface was not so steep as with thinner clothing. Skin temperature was higher and the subjects felt comfortable. However, this was compatible with a high rate of heat flow which caused the fall in body temperature. This did not arouse shivering since the relatively high skin temperature inhibited the shivering response. These experiments do not appear to have been repeated but the findings support the contention that heat flow rate as well as skin temperature need to be studied.

The deep-sea diver is often required to carry out tasks which appear to require quite modest levels of energy expenditure. However, increased heat production at great depths soon results in sweating and can easily lead to a rise in body temperature. The physiological mechanisms of temperature regulation are profoundly disturbed by the oxy-helium atmosphere and have to be implemented by a very careful control of the environmental temperature. Such control may become the dominant feature of body temperature regulation and it is the failure of such control which has led to a number of underwater accidents, some of them fatal[26].

The effect of exercise on body temperature at depth is presumably due to the failure of the sweating mechanism to balance the increased heat production. This in turn is due to the saturated atmosphere, hence the evaporation of sweat is limited. Skill and sophisticated equipment are necessary to control local conditions so that the diver is protected from thermal difficulties or even disasters.[27]

Heat loss in a helium atmosphere

As we have seen convective heat loss from the body to the surroundings is expressed by $Q = k\, dT/dY$ per unit area where k is the thermal conductivity of the fluid surrounding the body and dT/dY is the temperature gradient in the boundary layer flow at the skin surface. The steeper the temperature gradient, the greater the heat transfer. Convective temperature profiles may be plotted non-dimensionally so as to apply to the convective flow field at any height and temperature by substitution of specific values in the equation. For any particular gas this non-dimensional temperature graph is unique and is dependent on the Prandtl number. If the equations for heat transfer within the boundary layer are examined the convective coefficient for any particular gas can be expressed by

$$h_c = \left(\frac{k}{\sqrt{\rho g}} \right) \times \text{Constant}$$

When this is used to compare heat exchange rates in nitrogen and helium it is found that some 1.95 times more heat is lost in the helium environment at atmospheric pressure. Moreover it can also be shown that the convective coefficient will increase as the square root of the gas pressure.

For these reasons it is not surprising that subjects exposed to hyperbaric helium will require higher ambient temperatures for comfort in order to compensate for the extra heat loss.

These theoretical considerations were investigated experimentally using a surface plate calorimeter[28] attached to a hand when exposed to helium, nitrogen and air in a sealed perspex chamber and measuring directly the local

Table 8.1 Convective heat loss in helium, nitrogen and air.

Convection coefficient	Helium	Nitrogen	Air
h_c (W/m². °C)	3.2	1.8	1.7

convective heat loss. These results are shown in Table 8.1 where the experimental values were within 10 per cent of the theoretical calculations.

References

1.   Yoshimura, M. and Yoshimura, H. (1969). Cold tolerance and critical temperature of Japanese. *International Journal of Biometeorology* **13,** 163.
2.   Wyndham, C. H., Morrison, J. F., William, C. G., Bredall, G. A. G., Peter, J., Von Rahden, M., Von Rahden, E., Holdsworth, L. D., Van Graan, C. H., Van Rensburg, A. J. and Munro, A. (1964). Physiological reactions to cold of Caucasian females. *Journal of Applied Physiology* **19,** 877–81.
3.   Hensel, H. (1981). *Thermoreception and Temperature Regulation.* Monograph of the Physiological Society. Academic Press, London and New York.
4.   Slonim, A. D. (1952). *Fundamentals of the general ecological physiology of mammals.* Academic Press of USSR, Moscow and Leningrad.
5.   Uprus, V., Gaylor, G. B. and Carmichael, E. A. (1935). Clinical study with especial references to afferent and efferent pathways. *Brain* **58,** 220.
6.   Edholm, O. G., Fox, R. H. and Wolff, H. S. (1973). Body temperature during exercise and rest in cold and hot climates. *Archives des Sciences Physiologie* **27A,** 339–55.
7.   Pugh, L. G. C. E. (1964). Deaths from exposure on Four Inns Walking Competition, March 14–15, 1964. *Lancet* **i,** 1210–12.
8.   Pugh, L. G. C. E. (1966). Accidental hypothermia in walkers, climbers and campers; report to the Medical Commission on accident prevention. *British Medical Journal* **1,** 123–9.
9.   Pugh, L. G. C. E. (1966). Clothing insulation and accidental hypothermia. *Nature* **209,** 1281–5.
10.   Pugh, L. G. C. E. (1981). Personal communication.
11.   Fox, R. H., Woodward, Patricia, M., Exton-Smith, A. N., Green, M. F., Donnison, D. V. and Wicks, M. H. (1973). Body temperatures in the elderly: a national study of physiological, social and environmental conditions. *British Medical Journal* **1,** 200–206.
12.   Collins, K. J., Easton, J. C. and Exton-Smith, A. N. (1981). Shivering thermogenesis and vasomotor responses with convective cooling in the elderly. *Journal of Physiology* **320,** 76P.
13.   Golden, F. St. C. (1980). Problems of Immersion. *British Journal of Hospital Medicine* **23,** 371–3.
14.   Emslie-Smith, D., Lightbody, I. and Maclean, D. (1981). Conservative management of urban hypothermia. In: *Hypothermia—ashore and afloat,* pp. 147–50. Ed by J. M. Adam. Aberdeen University Press, Aberdeen.
15.   Ledingham, I. McA. (1981). Management of urban hypothermia: the aggressive approach. In: *Hypothermia—ashore and afloat,* pp. 151–7. Ed by J. M. Adam. Aberdeen University Press, Aberdeen.
16.   Collins, K. J. (1983). *Hypothermia—the facts.* Oxford University Press, Oxford, Toronto, New York.
17.   Ostrach, S. (1953). An analysis of laminar free convection flow and heat transfer about a flat plate parallel to the direction of the generating body force. NACA Rep. 1111.
18.   Pugh, L. G. C. E. and Edholm, O. G. (1955). The physiology of Channel swimmers. *Lancet* **ii,** 761–8.

19.   Wyndham, C. H., Williams, C. G. and Loots, H. (1968). Reactions to cold. *Journal of Applied Physiology* **24,** 282-7.
20.   Hayward, M. G. and Keatinge, W. R. (1981). Roles of subcutaneous fat and thermoregulatory reflexes in determining ability to stabilize body temperature in water. *Journal of Physiology* **320,** 229-51.
21.   Behnke, A. R. and Yazlou, C. P. (1950). Response of man to chilling and rewarming. *Journal of Applied Physiology* **3,** 591-6.
22.   Golden, F. St. C. and Hervey, G. R. (1977). The after drop of temperature following cooling. *Journal of Physiology* **272,** 26-7P.
23.   Burton, A. C. and Edholm, O. G. (1955). *Man in a Cold Environment.* Edward Arnold, London.
24.   Collins, K. J., Easton, J. C. and Exton-Smith, A. N. (1982). Body temperature after drop: a physical or physiological phenomenon? *Journal of Physiology* **328,** 72P.
25.   Elliott, D. H. (1981). Underwater physiology. In: *Principles and Practice of Human Physiology*, pp. 309-52. Ed by O. G. Edholm and J. S. Weiner. Academic Press, London and New York.
26.   Keatinge, W. R., Hayward, M. G. and McIver, N. K. I. (1980). Hypothermia during saturation diving in the North Sea. *British Medical Journal, 1,* 291.
27.   Bennett, P. B. and Elliot, D. H. (1982). *The Physiology and Medicine of Diving*, 3rd ed. Baillière Tindall, London.
28.   Toy, N. (1976). *Local free and forced convection heat transfer from the human body.* PhD Thesis. The City University, London.

9
Perception and Appreciation of the Thermal Environment

Thermal sensation

The way in which the central nervous system can appreciate the sensations of temperature and of heat and cold is dependent on the information received from the temperature receptors in the skin. These receptors are widely distributed in the superficial layers of the skin over the whole body surface but they are more concentrated in the extremities, in the fingers and toes and to a lesser extent in the hands and feet; the face is also very well supplied with temperature sensitive nerve endings. The endings themselves are probably undifferentiated; they have no very precise structure.

It has not been possible so far to isolate cold or hot endings to see what differences, if any, exist in their micro-structure. Their functional properties have been extensively investigated and can be quite precisely described. The temperature sensitive endings in the skin are stimulated by changes in temperature. When this happens a series of impulses is set up which travels along the afferent nerve and then passes to the posterior columns in the spinal cord and eventually reaches consciousness when the impulses arrive at the sensory cortex in the brain. *En route* these impulses also pass to the hypothalamus, which is the main centre for temperature regulation in the body.

This centre receives information from all of the body surface concerning temperature. The nerve endings can be divided conveniently into cold and hot endings. Cold endings are stimulated by a fall in temperature and hot endings by an increase in temperature. There is a family of nerve endings (as shown in Fig. 9.1) each having its own particular characteristics which determine at what temperature it will begin to discharge. One nerve ending may begin to discharge when the temperature drops from 25 to 24°C but will do so more vigorously below say 20°C; with a further fall in temperature this particular ending may discharge less frequently. Another ending may not start discharging until the temperature has dropped to say, 18°C and may reach its maximum firing rate at something like 12–13°C before diminishing. This pattern of activity is similar for all temperatures up to about 40–41°C[1].

Although detailed information is received at the regulating centre in the hypothalamus (and is essential for the control of body temperature) sensation is not aroused until the stimulus reaches the sensory cortex and we experience a feeling of heat or cold coincident with temperature change. Temperature sensors adapt quickly to a particular level of stimulation, and if there is no change in temperature the nerve endings concerned rapidly stop discharging. In a state of thermal comfort, we are not conscious of the temperature of our

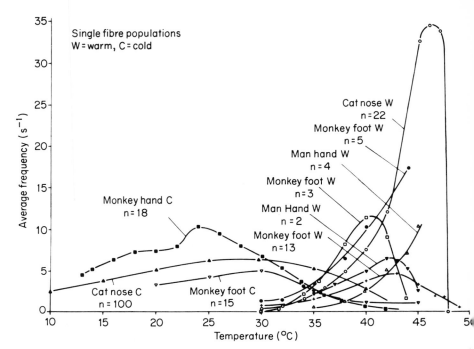

Fig. 9.1   Average static discharge frequency of populations of cold and warm nerve fibres as a function of skin temperature. (Reproduced with permission from Hensel (1981)[1].)

skin. We do not say to ourselves 'I feel comfortable' or 'my skin temperature feels comfortable'. Such thermal sensations cannot be regarded as being very precise in terms of absolute temperature. Our nerve endings do not act as thermometers but as indicators of change; either of increase or decrease. This is the basis for some of the paradoxical sensations that are so easy to demonstrate.

One such example of the difficulties encountered in interpreting thermal sensations is given here. The temperature of a subject's (A) knees were 32.0°C for the left knee and 30.6°C for the right. The subject assessed the left knee to be the coldest.

A second subject (B) reported that A's left knee felt hotter than the right. Subject A said that B's hand felt hot on the right knee and colder on the left knee. The palm temperature of subject B was 33.2°C. Subject A, on feeling his knees with his own right hand (palm temperature of 35.1°C) said that the left knee was slightly warmer than the right.

As far as A was concerned, the coolest knee had the highest temperature. But B with the direct contact assessment of the two knees said that the left knee seemed to be the hottest. The sensation of subject A was most probably determined by the quantity of heat lost from the knee to the environment and not by the absolute temperature of the skin surface. The assessment by subject

B appeared to be based on the temperature of the skin equated to the sensations of hot or cold, rather than to energy transfer. Such an observation raises questions about whether surface temperature or heat transfer to the environment is the dominant parameter in thermal sensation.

Thermal comfort

The subject of thermal comfort is not only of physiological interest but is of practical importance in specifying and providing satisfactory conditions inside the built environment. Architects, builders and designers need to know the conditions that the majority of human beings will find comfortable. The first problem is to design scales of warmth that can be used to investigate the sensations of people living or working in particular situations. It is only possible to find out how people are feeling, and whether they are comfortable or not, by questioning them; it is no use relying on theoretical considerations.

This problem was first tackled in the 1930s by Bedford[2] who, with his colleagues, visited many factories where men and women of different ages were employed. They made measurements of temperature, air movement, humidity and thermal radiation. Those individuals in the immediate vicinity of the measuring instruments were asked how they felt; did they feel comfortable, would they prefer to be warmer or cooler? They were also asked questions about sensations such as stuffiness. Gradually Bedford developed a simple scale (Table 9.1) and the questions became carefully standardized. He would ask whether an individual would prefer to be slightly warmer or slightly cooler. If a subject said no, that he did not want any change, then he was listed as being comfortable. There were many who would say that they were comfortable, but comfortably cool and others that they were comfortably warm. As the temperature departed further from these particular levels there would be an increasing number of people who would complain that they were too warm or too cool or that they were much too warm or cool.

With this scale Bedford and others investigated the preferred temperatures of people working under a variety of conditions and living in many parts of the world. Table 9.1 shows the Bedford scale, and a similar scale which was designed at ASHRAE, (the American Society of Heating, Refrigeration and Air-conditioning Engineers). This society maintains important laboratories at Pittsburgh where many studies of thermal comfort have been made.

Table 9.1 The 'Bedford' scale and the 'ASHRAE' scale of warmth.

The Bedford scale	The ASHRAE scale
Much too warm	Hot
Too warm	Warm
Comfortably warm	Slightly warm
Comfortable	Neutral
Comfortably cool	Slightly cool
Too cool	Cool
Much too cool	Cold

The range of temperature in which the majority of people feel comfortable was named the 'comfort zone' by Yaglou[3] one of the early workers in this field. He showed that it is impossible, except with an extremely wide range, to include everybody within this zone. There will always be some who feel too hot or too cold when for others the conditions are comfortable. This problem of variability, is one of the realities of the situation, in most people's experience.

The levels of temperature which are found within the comfort zone in the UK and USA are in the range 21–24°C provided that the air movement is no more than about 0.2 m/s and the relative humidity between 30 and 70 per cent. This zone may show seasonal differences, between summer and winter, and some differences have been found between the UK and USA although these are not large.

The concept of comfort is an abstract one which does not coincide with any specific physiological sensation. Thermal comfort may be better described as the absence of sensation or discomfort. In a state of thermal comfort, we are generally unaware of any temperature sensation and there are no parts of the body feeling either too hot or too cold. This is a difficult idea to grasp but it is quite clear that there are no specific comfort sensory nerves which are stimulated in a state of comfort to give rise to a positive sensation.

There are a number of physiological factors which affect thermal comfort. Our sensations are primarily determined by the skin and deep body temperatures, which are both sensed in the hypothalamus. It is here that the temperature regulating centre uses the information to control the various mechanisms of heat loss and heat gain as discussed in Chapters 3 and 4. Both skin and deep body temperatures give rise to sensations in the sensory cortex; however, we are not in general aware of our deep body temperature unless it is changing fairly rapidly, although this temperature comes into our overall appreciation of thermal sensation and comfort. This complex interaction can be described by experiments to determine the effects of body temperature on the peripheral circulation.

Subjects reclined in a well-stirred water bath; the unimmersed parts were wrapped in cotton wool so they were well insulated, leaving only the face exposed. If the water in such a bath is stirred rapidly, then the skin temperature and the water temperature are virtually identical[4]. In these experiments the bath temperature was at first kept at about 35.5°C which represented a neutral situation in which the subject was perfectly comfortable and would often go to sleep. The body temperature was then raised from 35.5°C to 41°C in 2 minutes by adding hot water into the bath. The subject immediately woke up feeling not only comfortable, but positively luxurious. The sensation of the hot water was at first extremely agreeable but gradually this feeling of comfort faded and was replaced by an increasing feeling of discomfort. During this period the water was kept at 41°C and the skin temperature was rapidly rising towards this value. At the same time body temperature began to rise, although this was slower than the increase of the skin temperature. As this happened, discomfort increased until by the time a body temperature of 38°C was reached the subject would start to complain and to become quite irritable. At this stage, the bath was cooled to about 38°C which made little difference to the feeling of discomfort. Deep body temperature then remained fairly constant between 38.1 and 38.3°C. For the next half hour the subject, in many cases, continued to be irascible. He was not only experiencing discomfort but his personality was to some extent changing for the worse. At the end of half an hour the bath temperature was lowered rapidly to 25°C and this cooling brought immediate relief and within a matter of seconds the subject would become his usual self. He was no longer cross or bad tempered and would feel extremely comfortable even though, at this point, the deep body temperature was still above 38°C and falling slowly to reach 37°C after some 10–20 minutes.

In this case the feelings of thermal comfort could be described as depending not only on skin temperature but also, to some extent, on deep body temperature. The most important element, if one could separate the two, would have been skin temperature. This was illustrated by the change in comfort sensation on cooling the bath temperature from 38°C to 25°C with the immediate restoration of comfort although body temperature was still raised.

Temperature sensation is not a good guide to the temperature of a particular part of the body, as illustrated by immersing the two hands in water at different temperatures; hot and cold. After a few minutes both hands are transferred to water at 30°C; one hand will feel cool and the other will feel warm although the actual temperature to which both are exposed is now identical.

Experiments carried out by Cabanac *et al.*[5] in France some years ago illustrate this point. He arranged for his subject to be riding a bicycle ergometer so that the heat production could be controlled either by sitting still or riding fast. Sitting at rest, lightly clothed in a room temperature of approximately 20°C, when the hand was put in water below 20°C it felt cold and unpleasant. The comfort temperature was over 38°C with 40–41°C being the preferred range. However, when the subject was cycling hard with a high rate of oxygen consumption, then the warm or hot water was disliked. It was only when the water temperature was 20°C or lower that the subject said this was no longer unpleasantly cold but was deliciously cool.

The results of this experiment show the importance of overall conditions in determining thermal comfort. The physiological factors of skin and deep body temperature have to be considered in conjunction with the influences of exercise, body size, body composition, sex, the effects of age and effects of adaptation or acclimatization to particular conditions.

Evidence will be presented first which appears to show that these factors are of considerable importance in the determination of thermal comfort. Evidence will then be given, which has been gathered in particular by Fanger and his colleagues, which show these factors have little or no effect on thermal comfort!

The comfort zone as worked out by Yaglou and Bedford was based upon surveys that they carried out under various conditions. Since that time a large number of surveys have been made all over the world in a variety of climatic zones, inside domestic dwellings, in factories, in offices, in hospitals, on board ship, in schools, etc. and usually either the Bedford or the ASHRAE scales have been used. Both of these seven point scales are relatively crude, in that there is a fairly wide temperature difference from one point on the scale to the next. Although the scales were not designed to be linear, it appears that the temperature differences between steps in the scales are of the order of 3°C and are more or less equal.

When using such scales the preferred temperatures, that is the temperatures where people claim that they do not want to be either cooler or hotter, show marked differences between different surveys. These surveys have been reviewed and summarized by Humphreys[6] who has shown that the range of preferred temperatures is from 17.1°C to as high as 31°C. These results are, to a large extent, attributed to different climatic conditions. People in hot countries prefer a higher temperature than those who live in cool or temperate regions. However, even within one particular climate, considerable variations are to be found. Originally there were differences shown by Bedford between men and women in which men appeared to prefer rather cooler temperatures than women. Since then, further surveys have not found such a large contrast and this has been attributed to changes in clothing habits and fashion, when men and women wear similar clothing there is still a difference in the preferred temperature, with women prefering slightly cooler temperatures than men. This has been assumed to be because they have a layer of subcutaneous fat approximately twice as thick as men. In addition, there have been marked effects attributed to age. On the one hand children who have a high metabolic rate do not appear to complain in environments where adults will say that it is too cold. At the other end of the scale there are large changes in temperature preferences with advancing age. The lowest preferred temperature quoted in the surveys reviewed by Humphreys was 17.1°C which was reported in a group of subjects over the age of 70, living in London, who were studied during the winter[7].

The effects of age, sex and body build (meaning the amount of subcutaneous fat) all appear to be quite well established as are the all important factors of clothing and activity. The metabolic rate is the most important physiological factor in determining the desired level of temperature for comfort. This is illustrated in Figure 9.2.

These findings regarding thermal comfort have been challenged by Fanger and his colleagues working in Copenhagen[8]. Subjects were examined under carefully controlled conditions in climatic chambers. They were all wearing similar clothing and all were sitting. They remained seated for periods of between 2 and 2.5 hours in the chamber, and were asked at frequent intervals if they wanted the temperature to be increased or decreased and whether they were comfortable. Temperatures were adjusted until the subjects claimed that they were entirely comfortable.

Under these conditions, Fanger has been unable to show any difference between old and young, between fat and thin or between men and women. He has been unable to show differences between people who have recently come from very hot climates and who might be considered to be acclimatized and those who have been exposed to cold conditions such as workers in a refrigerated store. He concludes that, providing conditions are properly standardized and controlled, all people have the same environmental requirements for their comfort.

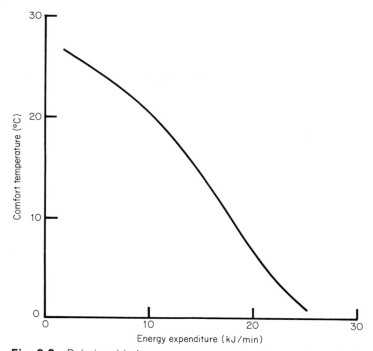

Fig. 9.2 Relationship between energy expenditure and comfort temperature for subjects standing and walking (at different speeds) out of doors and wearing one Clo.

In association with Nevins in the United States, Fanger carried out a wide range of studies with different activity and clothing conditions for a number of subjects. This resulted in the publication, in his book on thermal comfort[9], of tables showing the desired temperatures for particular levels of activity with various kinds of clothing. He has developed an equation combining all the factors concerned in thermal comfort, and this is as follows:

$$\frac{M}{A_{Du}}(1-\eta)-0.35 \left[43-0.061\frac{M}{A_{Du}}(1-\eta)-p_a\right] - 0.42 \left[\frac{M}{A_{Du}}(1-\eta)-50\right]$$

$$-0.0023 \frac{M}{A_{Du}}(44-p_a)- 0.0014 \frac{M}{A_{Du}}(34-t_a)$$

$$=3.4 \times 10^8 t_{cl}\left[(t_{cl}+273)^4-(t_{mrt}+273)^4\right]+t_{cl}h_c(t_{cl}-t_a)$$

A_{Du} = DuBois surface area; p_a = water vapour pressure; M = metabolic rate; t_a = air temperature; t_{cl} = clothing surface temperature; η = experimental mechanical efficiency; t_{mrt} = mean radiant temperature; h_c = convective cooling coefficient.

It is evident that there are fairly considerable differences both of finding, fact and of interpretation regarding the desired levels of temperature. These differences between, specifically, Fanger on the one hand and Humphreys on the other, have been discussed recently by MacIntyre[10] who points out that some of the variation which appears to be so important may be regarded as statistical. In numerous surveys where the seven point scale is used there is a high 'noise' level because of steps in the scale between points. So-called neutral temperature as determined with the seven point scale, and the preferred temperature as measured in a climatic chamber are not identical. Specifically, the neutral temperature can be biased by experience but does undoubtedly vary over a wide range.

MacIntyre has used another scale in which he asks the subjects; 'would you like to be warmer, have no change, be cooler?'. He has compared the answers obtained with this scale with those using the Bedford scale and shows, for example, that the optimum temperature occurs when curves of warmer request cross those of cooler request. This point is where the maximum number of people do not want any change in temperature. This should be at the neutral point of the warm scale, but is displaced upwards. MacIntyre emphasizes that in the winter people may vote 'no change' and although nobody would want the temperature to be cooler a number would like it to be a little warmer. On the other hand, in summer, people will state at about the neutral temperature that they do not want the conditions to be any warmer, but would like them to be a little cooler. Part of the difference is apparently verbal and depends on the interpretation by the subject of the words 'warm' and 'cool'. In a cold climate, people want to be warm and in a hot climate they want to be cool. There are very few people, for example in the winter in England, who would describe themselves as being comfortably cool. This may be related to the fact that when a subject is tested on a number of separate occasions in the field environment there is considerable variation in his responses, whereas when the same subject is in a climatic chamber his response is more consistent. Nevertheless, it is still difficult to completely reconcile these two sets of findings.

It may very well be that there is a difference in making a determination in one's own habitual environment compared with the rather unusual conditions of a climatic chamber. It is not customary for people to live under conditions which are strictly controlled. Indoor temperatures do not remain constant throughout the day and people are not exposed to exactly the same conditions for two or three hours at a time. They get up, walk about, go out of the room into perhaps a cooler or warmer corridor, they may not always sit in exactly the same position in the room and so on. For these reasons, the conditions are not strictly comparable between the chamber and the usual indoor climate, and this may be part of the explanation for the reported differences. It seems improbable that those who have gone to live in hot climates and who are adapted to these conditions would have exactly the same temperature preferences as those who are habituated to, or live in a very cold climate. It also seems extremely improbable that old people, in whom there are very obvious differences of temperature regulation, or more accurately a failure to some extent of temperature regulation, will have exactly the same thermal comfort votes as younger people.

Amongst the ways of explaining the differences between the findings of Fanger and his colleagues and those summarized by Humphreys and others are the ways in which the chamber tests are carried out. Auliciems[11, 12] has pointed out that in a situation where subjects are asked directly if they are comfortable, too cold, too warm, or if they prefer to have the temperature raised or lowered, it is almost impossible to avoid some degree of unconscious suggestion by the observer. This is such a well recognized effect that in nearly all experiments set up by psychologists considerable care is taken to avoid such bias. In clinical trials of drugs the double blind procedure has been adopted, where the choice of patient to receive a drug is determined by the use of tables of random numbers, and the drugs to be administered having a code unknown to those carrying out the tests. The doctors who perform the final clinical assessment have no knowledge of the drugs that the patients may have received. In this way it is possible to get a reasonably objective assessment of the therapeutic action of a particular drug. A similar rigorous procedure seems to be essential in order to make sure that the findings reported by Fanger and his colleagues can be substantiated.

There are a number of other points related to this question of suggestability. The ASHRAE scale defined thermal comfort as that condition of mind which expresses satisfaction with the thermal environment. This definition implies that there are not only physiological but also psychological aspects involved in thermal comfort. The expectation of the individual subject can play an important part and one which it is extremely difficult to control. It does not appear that the subjects' expectations have been assessed in any of the climatic chamber experiments reported so far. This criticism also applies to field studies where expectations may influence the result even more strongly than in the chamber. In a hot climate people expect the conditions to continue hot from day to day and do not have the expectation of a cool day; their psychological state is different from the person who is adapted to living in a cooler climate.

Finally, there is the question of personality which, although in itself is unlikely to modify in any way the individual's capacity to regulate temperature, nevertheless can affect the way in which the subject assesses his states of comfort. Again, there does not appear to be much investigation into the way that personality may affect assessment of conditions within a climatic chamber.

Some of these psychological points have been considered by a number of workers. Gagge, Stolwijk and Hardy[13] used three scales to describe pleasantness (ranging from unpleasant, slightly unpleasant and indifferent to pleasant); comfort (ranging from very uncomfortable, slightly uncomfortable to comfortable) and temperature (from hot to warm, to slightly warm, to neutral). By analysis of the responses of subjects to these scales they have shown that the minimal assessments are coincident, that is, that all three scales reduce to the lowest point at the same temperatures. However, as the ambient temperature diverges from the neutral value these three scales diverge from each other. In uncomfortably cold or unpleasant environments there were considerable differences in the assessments; a temperature of approximately 18°C was stated to be unpleasant but only slightly uncomfortable, the temperature itself being described as cool but not cold.

The words which are used to describe the thermal environment can be critical. The differences between groups of people may have been exaggerated by using the Bedford scale. On the other hand, it seems highly unlikely that Fanger's findings can be applied without modification to the field environment.

Indoor temperatures

It is important, when considering thermal comfort, to know what temperatures people maintain in their own homes. Are these inside conditions the best indicator of thermal comfort, or are they determined solely by economic factors, by the cost of heating or by the type of construction of the house; how far do these factors interact?

There have been a number of reports by government commissions dealing with temperatures inside domestic dwellings. In 1945 recommended temperatures were, that bedrooms should be kept between 50 and 55°F (10–13°C), kitchens 60°F (16°C) and living rooms 58–62°F (14.5–17°C). Since then there has been a gradual rise: in 1961 the Parker–Morris Committee recommended that living room temperatures should not fall below 65°F (18.3°C) and bedrooms should not be colder than 50°F (10°C) when outside temperatures were 30°F (−1°C).

By 1977 the British Standards Institution gave design temperatures of 70°F (21°C) for living rooms, and 65°F (18.3°C) for bedrooms, kitchen, toilet, when outside temperatures were 28°F (−2°C).

Measurements of temperatures inside houses, flats and bedsitting rooms have been made by a number of investigators; the most extensive studies being those carried out for the UK Electricity and Gas Councils and by the Building Research Station. Many of these surveys have been concerned with comparisons of various heating systems in terms of energy consumption and cost. Recently, D. R. G. Hunt[14] of the Building Research Station has reviewed the measurements that have been made and has been able to detail observations made in over 3500 homes, and to show that there has been an increase of indoor temperatures in recent years at the rate of 1°C per decade. It appears that this upward trend has halted at about 21°C. In a personal communication Hunt has given some preliminary information about the results of a survey in 1000 homes distributed throughout the UK, and including a variety of types of dwellings, but the large quantity of data has not yet been fully analysed. In January, 1978, with a mean outside temperature of 7.5°C, the mean daytime temperature in 'downstairs' rooms (living rooms, kitchen, hall) was 16.5°C and 14.0°C upstairs in bedrooms and bathrooms.

Fox *et al.*,[7] during a study of hypothermia in the elderly, measured living and bedroom temperatures on one day in the morning and evening in the homes of over 1000 subjects. The measurements were made from January to March when the outside temperature averaged 4.9°C. The morning temperature in the living room averaged 17.5°C, and in the evening 19.0°C. Bedroom temperatures were about 1.5°C lower; but the range was large for both rooms being from 5 to 25°C. About a quarter of the 1000 subjects slept in bedrooms with temperatures below 14°C. Ten per cent of the morning living room temperatures were below 12°C. Such temperatures are not only to be found in the homes of the elderly. Of the various studies included in Hunt's survey, in 10 out of 36 surveys the living room temperature was below 18.3°C (65°F) in the evening.

A study by the Electricity Council of 69 homes in 11 different sites in England, ranging from Gateshead to Plymouth, carried out from 1967 to 1973, showed that in a cold week 35 out of 69 living rooms and 57 out of 69 bedrooms were below 18.3°C.

In spite of the overall indoor temperature increases since 1945, many homes are kept at temperature levels which are generally considered to be too low. These various surveys have been reviewed by Edholm and Lobstein[15].

We need to know more about indoor temperature if we are to consider the question of conservation of energy by reducing the amount of power used in houses. With the rise in indoor temperatures in the UK since the Second World War there have been changes of clothing habits, with a reduction of insulation particularly by men.

Temperature and performance

Recently Nichol has made a study of telephonists to examine the relationship between air temperatures and the speed and accuracy of the work of the telephone operator. Nichol's findings, which have not been published, showed that there were changes in peformance with temperature.

The number of errors went up sharply, and delays increased, as the temperature fell below about 24°C. There was little effect on measured performance if the temperature rose above 24°C to 25–26°C.

Fox[16] studied the errors in typing and the effect of the temperature of the hands and arms (Figure 9.3). Fox's subject immersed her arms in water of different temperatures on different occasions for a period of approximately 20 minutes. The arms were then dried, and the subject carried out a standard typing task whilst the speed of typing and the errors made were measured. Immersion in water at temperatures above 34°C, that is 36–38°C made little difference to performance. When the water temperature was reduced below 34°C there was a decline in typing speed, and at 33°C an increase in the number of errors which became progressively greater with a further fall in temperature. It appears from these two studies that environmental temperature and the local temperature of hands and arms, can have an important effect upon performance.

These changes in skill could not be due entirely to variations in the skin temperature of the fingers. There was no indication of any significant loss of sensation occurring above about 30°C, and it is much more likely that the effect was due to the cooling of the muscles. It has been shown by Barcroft and Edholm[17] that when the arms are immersed in cold water (without any change in the environmental conditions of the body as a whole) the subcutaneous tissue and muscle temperatures fall as illustrated in Figure 9.4. Such muscle cooling and an associated increase in viscosity, together with changes in conduction times in the motor nerves, seems to be the most likely reason for slowness of movement and decreasing skill at low temperatures.

From the point of view of the conservation of heat energy such findings are

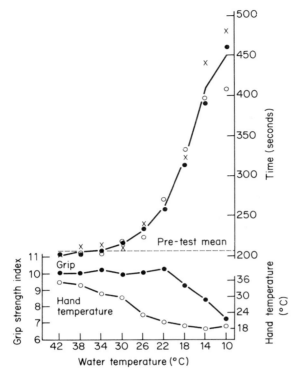

Fig. 9.3 The effect on copy typing speed of immersing both hands and forearms for half an hour in water at temperatures ranging from 10 to 42°C. The top curve shows the time taken to complete a standard typing task. Performance becomes slower after immersion in water at 30°C or below. *Principles and Practice of Human Physiology.* Academic Press. (Reproduced with permission from Edholm, O.G. and Weiner, J.S. (1981).)

perhaps unfortunate, as they indicate that there would possibly be serious effects from reducing indoor temperatures much below the current figures, particularly regarding offices and factories. In the home it may be that domestic skills would not be as sensitive as the skills of typing. It appears there is little potential for useful conservation of heat by reducing present temperature levels. There is the alternative of reducing heat loss, not only from the house, but also from the individual by increasing the insulation of clothing. It is evident from examining current fashions that the majority of people today have much lower insulation in their clothing than was customary even 25 years ago. The obvious suggestion could therefore be made that if it is desired to save fuel then people should be advised to wear thicker clothing.

Effects of cold
The temperature of the environment can affect the performance of tasks in two ways; firstly, the physical demand of a cold environment requires a higher

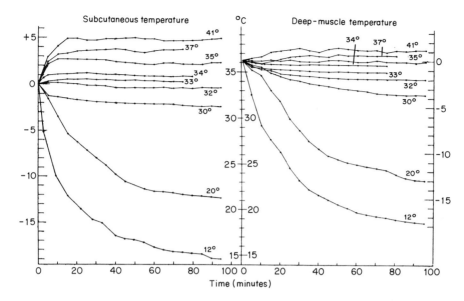

Fig. 9.4 The subcutaneous and deep muscle temperatures in a forearm placed in a water bath at time 0. The water bath temperature is given above the right hand end of each curve. In the centre of the figure the ordinates show the subcutaneous and deep muscle temperatures in °C. On the right and left of the figure the ordinates are given as deviations, in °C, from the average temperature in the clothed arm; on the left side 0 = 33.6°C, the average subcutaneous temperature in the clothed forearm; on the right 0 = 36.2°C, the average deep muscle temperature. (Reproduced with permission from the *Journal of Physiology*)[17].

energy expenditure for a given amount of work, and secondly the effect of temperature may be more basic by acting directly on the nerve endings in the skin and changing the conduction rate in the nerve fibres. A fall in ambient temperature causes movements to become slow and clumsy because of the increased viscosity associated with the cooling of muscles and joints. In the cold, energy expenditure may be high owing to an increase in metabolic rate and shivering, but can also be due to physical conditions of work being much harder. The energy costs of walking can be considerably increased when moving on ice or snow or a mixture of the two. Brotherhood[18] investigated the effects of antarctic conditions on walking and the energy cost compared with walking at a similar speed on a uniform surface in a temperate climate; the comparison varied some two-threefold.

Another way in which environmental conditions affect the cost of work is found in antarctic or arctic conditions when stores have to be dug out of the snow. Such work can involve a considerable energy expenditure compared with doing the same task in a moderate climate when boxes of food are not buried. A more important effect can be seen when carrying out the work of repairing or maintaining a mechanical vehicle when the increased viscosity of lubricating oils makes the energy cost of undoing nuts and bolts considerably higher. Not only is the task itself harder but it will take longer: it has been estimated that to maintain a vehicle in the antarctic takes anything up to 20 times longer than when this is carried out in comfortable temperate conditions. Another factor is the hampering effect of garments; with thick polar clothing movements tend to become clumsy and the clothes themselves impair arm and leg movements. However, the effect on energy expenditure appears to be quite small[19].

Apart from these physical or mechanical disadvantages of working in cold conditions there are other ways in which the temperature can affect the capacity for work. Mackworth[20] demonstrated that the degree of finger numbness due to cold exposure is less in the acclimatized than in the non-acclimatized individual. Finger numbness may be measured by determining the size of a gap between two points before they are identified as separate. With a warm finger, these points may be very close together, about 1 mm apart, and will be sensed individually; in a cold finger they may have to be 1 cm apart before they are determined as discrete points. As numbness increases, the ability to carry out fine movements decreases. With completely numb fingers, it is extremely difficult to carry out most skilled manipulative tasks.

In general, with adequate polar clothing, and with the additional physiological reactions of peripheral vaso-constriction and shivering, body temperature does not usually fall, but with continued or enforced outdoor exposure it may begin to drop; this has some effect on skills and ability presumably due directly to a change of body temperature on the brain and on the nerve centres within the brain.

Effects of heat
The effects of heat on performance are considerable. In hot conditions, the body temperature can rise and the individual may experience discomfort from increased skin temperature and sweating, and this can affect the performance of some tasks. In many industries workers can be exposed to severe heat, usually in the form of radiation, or so-called 'wild' heat. In steel works there can be intense radiation from molten metal, and when furnaces have to be repaired heat loads can be severe. In the ceramics industry hot ovens are used for baking and in glassworks molten glass can provide high radiant heat loads. It is of some importance to know the safe tolerance times of individuals working in severe heat conditions. These have been determined by Bell *et al.*[21] and Figure 9.5 can be used for determining the length of time for which people may be safely exposed without a risk of collapse. Bell *et al.* found that the best way of combining wet and dry bulb temperatures was by the formula $T = 0.23\,DB + 0.77\,WB$. The effect of wet bulb temperature is to prevent, or diminish, heat loss by the evaporation of sweat; high wet bulb temperatures impose a severe constraint on human work. There are marked individual variations in heat tolerance as one example given by Bell *et al.* shows. Tolerance times at a temperature of 40°C DB, saturated with water vapour, were 16 minutes for 99 per cent of the subjects; 95 per cent of the subjects would have been safe for 20 minutes, and 50 per cent would have been able to work

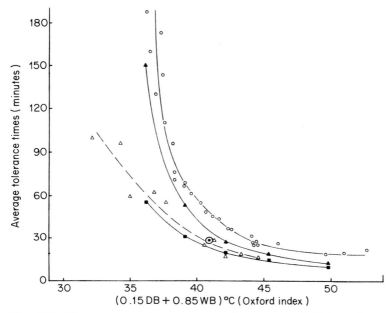

Fig. 9.5   The average tolerance times of men, seated (○), and at energy expenditures of 145 W (▲), 210 W (△) and 325 W (■). The symbol ◉ represents the average tolerance times of seated men when the air movement was 4.5 m/s; the remaining results for seated men are drawn from conditions where the air movement did not exceed 1.5 m/s. 'Safe' tolerance times should be taken to be 75 per cent of the average time shown in this figure. Adapted from Lind (1963) *Temperature, its measurement in science and industry*, Vol. 3, Part III; Reinhold Publishing Corporation.

for 22 minutes. The reasons for this variability are not clear, although they have been noted by everyone studying the effects of heat on man. One significant factor is body weight, or more precisely, subcutaneous fat thickness. The greater the fat thickness the more likely is the individual to collapse under hot conditions, but this is not the only factor. In the work of Bell *et al.* the subjects were unacclimatized to heat, whereas those people who work in some industrial conditions may become well acclimatized.

Tolerance times represent the extremes of the effects of heat on performance. Mackworth[22], in studies carried out in Singapore, examined the way that various environmental conditions affected the performance of 'wireless telegraphists'. He showed that the initial degree of skill played a considerable part. Subjects who had only just been recruited into the service were compared with those who had a long experience and were considered first class at the job; he also had a group with intermediate skill. At comfortable temperatures the performance of these groups was not very different; however, when the temperature was raised the least skilled people began to make more mistakes at a time when the skilled and semi-skilled were able to maintain a normal standard. With further increases in temperature, both the semi-skilled and the unskilled showed marked deterioration, but temperatures had to be taken higher still before the skilled workers showed any decline. In general it was considered that an effective temperature of about 30°C (86°F) was the point at which deterioration in performance set in; this occurred without any changes in body temperature. In a series of experiments carried out by Wilkinson, *et al.*[23], the effect on performance of raising and holding body temperatures at particular levels was studied. The tasks used were arithmetic and auditory vigilance tests. As body temperature was raised, errors in arithmetic increased; the higher the body temperature the greater this deterioration. On the other hand, with the auditory vigilance test more signals were detected as body temperatures increased. It is dangerous to assume that a rise in body temperature will produce a deterioration in all tasks. It has been suggested that the reason for the better performance in the vigilance test at the higher body temperatures was due to a greater level of arousal. This finding has also been confirmed by Colquohoun[24] who studied a visual vigilance task and found improvement with a raised body temperature. This cannot be taken too far; the levels of body temperature studied by Wilkinson, *et al.*[23] were 37–38.5°C. If the temperature had been raised much above 39°C it is probable that there would have been a deterioration in all tasks, including vigilance.

Another effect of high environmental temperatures is that as work continues the sweat rate decreases. The so-called sweat gland fatigue is initiated very early with exposure to hot conditions; peak sweat rate is reached as soon as body temperature has risen. Thereafter, sweat rate tends to decline; if work continues there may be a further rise in body temperature and a deterioration in performance with the worker becoming a possible candidate for heat illness including heat cramps (which may be relieved by salt) and cardiovascular collapse. The worker who continues in hot conditions when his sweat rate is falling is liable to the more serious conditions of heat stroke; sweating can then be inhibited and the individual may have a dry skin with a rapidly rising body temperature. Heat stroke is a dangerous condition and once body temperature exceeds approximately 42°C there is always the possibility of death; when temperatures reach 43°C or more, permanent damage can occur to the brain; recovery from body temperatures of 44–45°C is rare indeed.

Apart from the effects of heat illness and collapse the experiments in Singapore showed that performance can fall off even without a rise in body temperature. This effect is not so easy to explain; it may well be that the increasing discomfort distracts attention from the work. Physical work is limited by environmental temperatures; work which can be carried out in comfort in temperate conditions becomes very severe in a hot climate. In hot countries, the rate of work is slower than in temperate areas. As an illustration, the speed with which people walk appears to be related to the environmental temperature. In hot weather people stroll rather than walk, and in hotter conditions they will avoid strolling and will preferably stand or, better still, sit or lie down. In many tropical parts of the world work is confined to the early hours and to the late afternoon, when temperatures will be considerably lower than at midday. This influence of the climate on the pattern of work differs in many parts of the world. In Israel, in the Negev region, temperatures during the summer months are high, of the order of 35–38°C with the wet bulb around 27–30°C. There is usually a moderate wind averaging about 1 m/s. The local farmers continuously sow and reap crops and in order to get a reasonable return for their work two, and sometimes three, crops are grown each year. Some of the work is as heavy in the summer as in the winter and is maintained throughout the day, beginning at approximately 06.00 and continuing until sunset, around 18.00 in the evening. Edholm *et al.*[25] observed that if there was a group of approximately six men engaged in raising a particular crop, they would work for a period of about three minutes and then pause, talk to their neighbour for perhaps 30 seconds, and then resume work. This spacing of work and rest periods was scarcely noticeable until careful timings were made. This regimen was effective, body temperature did not rise and heart rate did not exceed about 100 beats a minute, the same as would have been expected for workers in cool or temperate areas. There was undoubtedly a considerable learning factor in overcoming these climatic conditions. Similar patterns of work have been observed in other countries; in New Guinea for example, Budd *et al.*[26] made observations of men and women in the island of Karkar, close to the equator, and also on highlanders in the village of Kaul, some 5000 feet above sea level. The villagers of Karkar were living in a hot and very humid climate and grew a variety of crops. Both men and women worked throughout the day for about ten hours starting between 06.00 and 07.00 in the morning and finishing between 16.00 and 17.00. They carried out a variety of agricultural tasks with frequent but short pauses in their work. None of the islanders was found to have high sweat rates compared to the Europeans; heart rates did not go up much beyond 110–120 beats a minute. It is difficult to understand how such work was continued except on the basis of conservation of energy by frequent rest pauses. In many other countries with a hot, humid climate it is customary for there to be a rest or siesta in the middle of the day with work continuing until fairly late in the evening. This may well be a valuable cultural habit; in those countries with marked seasonal differences, people will take a siesta during the summer but are liable to continue to do this even when cooler weather has arrived. There is a good deal of habit and custom involved rather than a physiological requirement.

Effect of body temperature on the appreciation of time

Fox *et al.*[27] studied the effect of changing body temperature on time sense. Figure 9.6 shows the results of their work: as body temperature rose, time appeared to pass more rapidly as judged by the way different subjects estimated the lapse of specific time intervals.

Heat stress indices

There have been many attempts to combine the different characteristics of a thermal environment into one overall index. This is difficult but the human body does it all the time. Man does not sense the environment in terms of humidity, dry bulb temperature, or of air movement except in a very rough

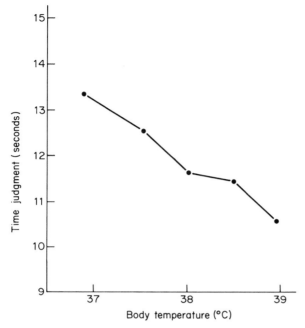

Fig. 9.6   Time judgement (10 s by counting) by a group of 12 subjects at different body temperatures.[27]

way. The overall effects are epitomized by him with a combination of different factors in the environment acting together to produce different quantities of heat loss or heat gain, which in turn give rise to the sensations of thermal comfort or discomfort.

The various indices which have been designed to describe particular conditions are based either on direct measurement of the environment or of the body's response to it. Measurements of dry bulb and wet bulb temperature, air movement, radiation and other environmental factors can be combined empirically into an index which can be tested by asking people to experience various combinations of temperature in order to decide if they produce the same kind of sensation. Most of the scales that are described here have been designed in terms of heat sensation rather than in terms of cold sensation.

Effective temperature (ET)
One of the first practical scales was designed by Yaglou in the 1920s and is called the Effective Temperature Scale[28, 29]. The effective temperature of an environment is the temperature of still, unsaturated air which gives rise to an equivalent sensation. In making this scale subjects were exposed to a particular combination of dry bulb and wet bulb temperatures in a climatic chamber, and they were then taken rapidly to a second room where the combination of dry and wet bulb was so designed that it would give approximately the same sensation. The subjects were asked if the environments felt

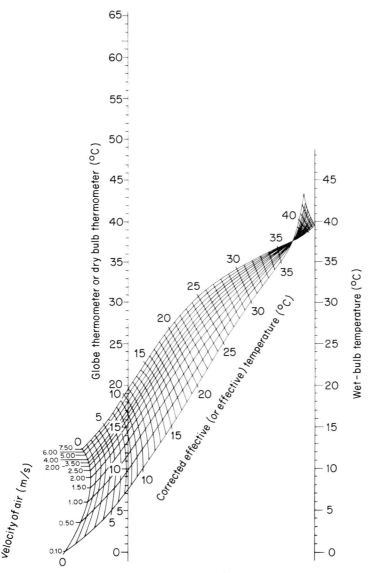

Fig. 9.7 Normal scale of corrected effective (or effective) temperature, used for subjects engaged in sedentary or very light work and wearing light indoor clothing. For effective temperature use dry bulb temperature. For corrected effective temperature use globe thermometer temperature. (From Ellis, *et al.* (1972) *British Journal of Industrial Medicine* **29,** 361–74.)

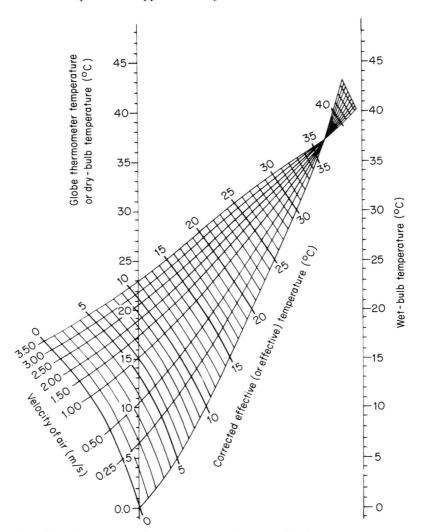

Fig. 9.8 Basic scale of corrected effective (or effective) temperature, used for subjects stripped to the waist. (From Ellis, *et al.* (1972) *British Journal of Industrial Medicine* **29**, 361–74).

similar or not and in this way, by using a wide variety of temperatures, Yaglou was able to construct a scale showing equivalence in environments of widely differing dry and wet bulb combinations.

The effective temperature scale was modified some years later by Bedford when he included a factor for radiation. This corrected effective temperature scale is still widely used and is valuable in moderately to fairly warm environments (e.g., 28–40°C). It is less useful in hotter conditions, and cannot be used at temperatures below about 20°C (Figs 9.7 and 9.8).

Wet bulb globe thermometer (WBGT) index
Perhaps the next most popular scale is the WBGT or wet bulb globe thermometer index. This was also designed by Yaglou together with Minard[30]
and was intended as a simple scale to be used out of doors. The original
purpose was to enable military officers to judge whether the environment was
too severe for military exercises. It was used initially at the marine corps
training camp in South Carolina, which has a very hot and humid climate,
and where recruits had to carry out extremely arduous training. This scale
proved to be effective; by providing specific information about conditions,
the number of cases of heat illness was greatly reduced. The number of hours
which were lost in training was rather less than it had been previously, when
it was purely a question of individual judgement as to whether the conditions
were too hot for particular exercises. The WBGT scale combines dry and wet
bulb temperatures and includes a measure of radiation by the use of the globe
thermometer (previously described). The WBGT includes a factor for forced
convective heat loss, or cooling due to the wind and the formula used to
calculate the temperature is:

$$0.2T_{globe} + 0.1T_{db} + 0.7T_{wb}$$

This is a simple scale, easily measured, and is useful out of doors, particularly
in conditions of considerable solar radiation. It becomes ineffective if dry
bulb temperatures are in excess of 40°C.

Belding and Hatch index
The third scale which can be used indoors, particularly in industry, is the
Belding and Hatch index[31]. This can be described by reference to Figure 9.9.
The various components of the thermal environment have to be measured in
order to construct the index stage by stage, starting with measurements of
radiation and convection, then including metabolic heat production of the
individual and finally determining the evaporative capacity of the environment.

This index is effective over a wide range of temperatures, the only problem
being that it is complicated to evaluate. For this reason it has not been so
widely used in industry as it should have been.

The predicted 4-hour sweat rate (P4SR) index
The predicted 4-hour sweat rate, or P4SR, is a scale based on the physiological
responses of the individual. It was developed from many experiments carried
out in climatic chambers by MacArdle and his colleagues during World War
II[32]. From measurements of the sweat rate of subjects wearing particular
clothing, and doing specified work over a 4-hour period, the nomograms
shown in Figure 9.10 were constructed. From this empirical approach, it is
possible, by measurement of the dry and wet bulb temperatures, together
with air movement, to predict 4-hour sweat rate. Corrections can be made for
the metabolic activity of the individual and there is also an adjustment for
clothing. The scale is useful particularly in very hot and humid conditions,
but there are some limitations which are due to the way the scale was
determined. The results were obtained on fit young men who were acclima-

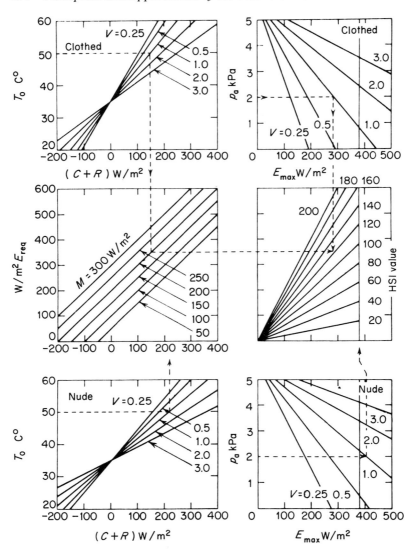

Fig. 9.9   The Belding-Hatch heat stress index. $(C+R)$ W/m^2 = (convection + radiation) T_o°C = operative temperature, p_a = ambient water vapour pressure, E_{max} W/m^2 = maximum evaporative capacity, W/m^2 E_{req} = evaporation required, M = metabolism (W/m^2). An example can be worked out by following the dotted line. (Reproduced with permission from Kerslake (1972). *The Stress of Hot Environments.* Cambridge University Press, Cambridge.)

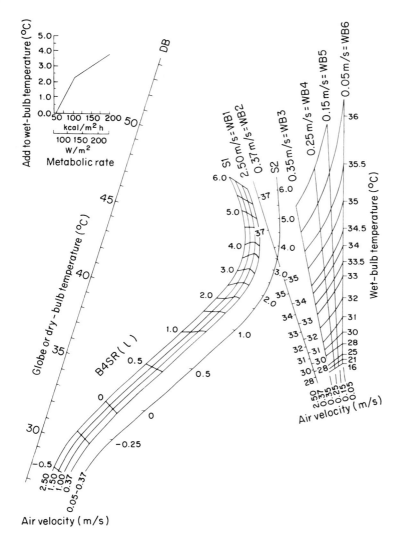

Fig. 9.10 The calculation of P4SR involves first the determination of the basic 4-hour sweat rate (B4SR) by drawing a line from the appropriate point on the temperature scale to the wet bulb temperature scale (using observed air velocity scale). The P4SR is indicated where this line intersects the B4SR scale at the given airspeed. The insert shows a nomogram for energy expenditure above resting. (Reproduced with permission from Kerslake (1972). *The Stress of Hot Environments*. Cambridge University Press, Cambridge.)

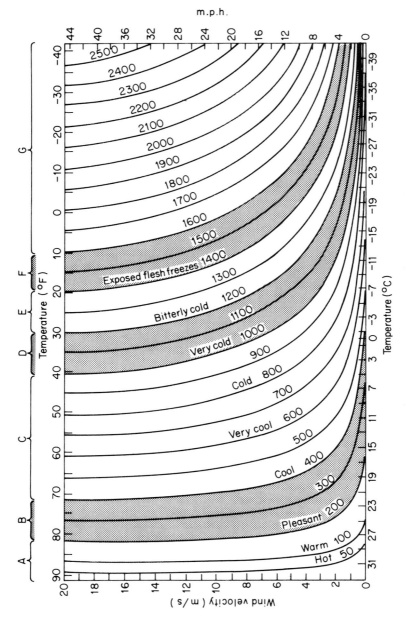

Fig. 9.11 Wind chill index, showing rate of cooling in kcal/m²/h at different combinations of wind velocity and temperature. The critical effect of wind velocity is evident. (Amended from the General Staff Training Publication, reproduced with permission of the Ministry of Defence.)

tized to heat. The sweat rates are higher than would be expected for unacclimatized people and also for those not so physically fit. Nevertheless, in spite of this, experience has shown that the P4SR is a reliable index and one which has been used in a wide variety of situations with many different subjects.

Wind chill index
The wind chill index or scale was originally designed by Paul Siple[33, 34] from observations made in the antarctic of the times taken for water to freeze in various climatic conditions (Figure 9.11). The important point about this scale is that it emphasizes the effect of air movement on cooling. If the air is still, low temperatures, of the order of -10 to $-40°C$, can be tolerated without much difficulty. If air movement increases, even by a small amount, then the cooling is greatly increased, as will be evident from the considerations in Chapter 4 of the results of forced convection on heat loss. This is clearly illustrated in the wind chill index and it is perhaps for this reason that it has continued to be so widely used 30 years after it was originally defined even though there are many criticisms both on theoretical and even practical grounds. For instance, in the descriptions given in Figure 9.11 the speed with which exposed flesh will freeze are somewhat exaggerated; the cooling time at the upper end of the scale is not as short as indicated. In spite of such criticisms, this scale still remains the most useful method of evaluating cold climates and cold conditions generally.

References

1. Hensel, H. (1981). *Thermoreception and Temperature Regulation.* Academic Press, London and New York.
2. Bedford, T. (1936). Warmth and comfort. *Journal of the Institute of Heating and Ventilating Engineers* **4**, 383–96.
3. Yaglou, C. P. (1926). Comfort zone for men at rest. *Journal of Industrial Hygiene* **8**, 5–16.
4. Edholm, O. G., Fox, R. H. and Wolff, H. S. (1973). Body temperature during exercise and rest in cold and hot climates. *Archives des Sciences Physiologiques* **27A**, 339–55.
5. Cabanac, M., Hildebrandt, G., Massonet, B. and Strempel, H. (1976). A study of the nycthemeral cycle of behavioural temperature regulations in men. *Journal of Physiology* **257**, 275–91.
6. Humphreys, M. A. (1975). *Field studies in thermal comfort compared and applied.* Building Research Establishment Current Paper 76/75.
7. Fox, R. H., Woodward, P. M., Exton-Smith, A. N., Green, M. F., Donnison, D. V. and Wicks, M. H. (1973). Body temperatures in the elderly: a National Study of Physiological, Social and Environmental conditions. *British Medical Journal* **1**, 200–206.
8. Fanger, P. O. (1973). The influence of age, sex, adaptation, season and circadian rhythm on thermal comfort criterion for man. *Annexe au Bulletin de I'Institut International du Froid* **2**, 91–7.
9. Fanger, P. O. (1970). *Thermal comfort.* Danish Technical Press, Copenhagen.
10. McIntyre, D. A. (1981). Design requirements for a comfortable environment. In: *Bioengineering, Thermal Physiology and Comfort.* Ed by K. Cena and J. A. Clark. Elsevier, Amsterdam.

11.   Auliciems, A. (1972). *The Atmospheric Environment. A study of comfort and performance.* University of Toronto, Department of Geography Research Publication. University of Toronto Press.
12.   Auliciems, A. (1981). Towards a Psycho-Physiological model of Thermal Perception. *International Journal of Biometeorology* **25**, 109–22.
13.   Gagge, A. P., Stolwijk, J. A. J. and Hardy, J. D. (1967). Comfort and thermal sensation and associated physiological responses at various ambient temperatures. *Environmental Research* **1**, 1–20.
14.   Hunt, D. R. G. and Steele, M. R. (1980). Domestic temperature trends. Personal communication.
15.   Edholm, O. G. and Lobstein, T. (1982). Indoor temperature. In: *Hypothermia; ashore and afloat.* Ed by J. M. Adam. Aberdeen University Press, Aberdeen.
16.   Fox, R. H. (1965). *Thermal comfort in industry. Ergonomics for Industry: 8 (Ministry of Technology).* HMSO, London.
17.   Barcroft, H. and Edholm, O. G. (1945). Temperature and blood flow in the human forearm. *Journal of Physiology* **104**, 366–76.
18.   Brotherhood, J. R. (1973). Studies in energy expenditure in the Antarctic. In: *Polar Human Biology.* Ed by O. G. Edholm and E. K. E. Gunderson. Heinemann Medical, London.
19.   Teitelbaum, A. and Goldman, R. F. (1972). The energy cost of clothing. *Journal of Applied Physiology* **32**, 743–4.
20.   Mackworth, N. H. (1953). Finger numbness in very cold winds. *Journal of Applied Physiology* **5**, 533–7.
21.   Bell, C. R., Hellon, R. F., Hiorns, R. W., Nicol, P. B. and Provins, K. A. (1965). Safe exposure of men to severe heat. *Journal of Applied Physiology* **20**, 288–92.
22.   Mackworth, N. H. (1950). Researches on the measurement of human performance. *MRC Special Report Series No. 268.* HMSO, London.
23.   Wilkinson, R. T., Fox, R. H., Goldsmith, R., Hampton, I. F. G. and Lewis, H. E. (1964). Psychological and Physiological responses to raised body temperature. *Journal of Applied Physiology* **19**, 287–91.
24.   Colquohoun, W. P. (1971). Circadian variations in mental efficiency. In: *Biological rhythms and Human Performance*, pp. 39–107. Ed by W. P. Colquohoun. Academic Press, London and New York.
25.   Edholm, O. G., Humphrey, S., Lourie, J. A., Tredre, B. E. and Brotherhood, J. R. (1973). Energy expenditure and climatic exposure of Yemenite and Kurdish Jews in Israel. *Philosophical Transactions of the Royal Society of London Series B* **266**, 127–40.
26.   Budd, G. M., Fox, R. H., Hendric, A. C. and Hicks, K. E. (1974). A field survey of thermal stress in New Guinea villagers. *Philosophical Transactions of the Royal Society of London Series B* **268**, 393–400.
27.   Fox, R. H., Bradbury, Pamela A. and Hampton, I. F. G. (1967). Time judgement and Body Temperature. *Journal of Experimental Psychology* **75**, 88–96.
28.   Yaglou, C. P. (1927). Temperature, humidity and air movement in industries; the effective temperature index. *Journal of Industrial Hygiene* **9**, 297–309.
29.   Yaglou, C. P. (1947). A method for improving the effective temperature index. *Transactions of the American Society of Heating and Ventilating Engineers* **53**, 307–13.
30.   Yaglou, C. P. and Minard, D. (1957). Control of heat casualties at military training centres. *American Medical Association's Archives of Industrial Health* **16**, 302–16.
31.   Belding, H. S. and Hatch, T. F. (1955). Index for evaluating heat stress in terms of the resulting physiological strain. *Heating, Piping and Air-conditioning* **27**, 129–36.
32.   McArdle, B., Dunham, W., Holling, H. E., Ladell, W. S. S., Scott, J. W., Thomson, M. L. and Weiner, J. S. (1947). *The prediction of the physiological effects of warm and hot environments. Ref. No. R.N.P. 47/391.* MRC, London.

33. Siple, P. A. and Passel, C. F. (1945). Dry atmospheric cooling in sub-freezing temperatures. *Proceedings of the American Philosophical Society* **89,** 177–99.
34. Court, A. (1948). Wind Chill. *Bulletin of the American Meteorological Society* **29,** 487–98.

10
Clothing

Principles of clothing

A significant cultural development by man has been his use of clothing, particularly in cold or cool climates to provide extra insulation against heat loss. The primary rule regarding clothing design in these conditions is the provision of still or dead air trapped in the fibres of the fabric. As has been frequently stated, the most important aphorism about the physiology of clothing is that the thermal insulation is proportional to the thickness of the dead air enclosed[1]. This principle was first derived during the Second World War when it became necessary to provide adequate insulation for many thousands of men exposed to severe, or moderately severe cold conditions. It had not been appreciated before then that the essential aspect was the amount of air trapped in the garment. Still or dead air—that is, air without any movement—is a very effective insulator, and it does not matter what kind of fibre is used to trap the air. Research, which was originally to discover fibres that might provide better insulation, switched to making sure that clothing could be made which would maintain a particular thickness of the air layer. The same principle applies to fur-bearing animals, where the hairs of the fur trap the air to provide an insulating layer. A fur coat might be described as consisting of very expensive air trapped betwen hair. The characteristic of air, that of low conductivity, means that the way in which fibres trap air must be in relatively small cells. Once a cell is of the order of $0.5\,cm^3$ there is the possibility of thermal currents existing within the air volume. With smaller interstices there is virtually no movement at all, and hence the effect is one of pure insulation by the air. The trapped air must remain immobilized in spite of movements by the wearer or due to wind penetrating the clothing and disturbing the layer. There is also the so-called 'bellows' action with movement, especially when several layers of clothing are worn; air is expelled from between the layers.

The consequence of these principles is that the density of the fibres should be low, ensuring that the whole garment is as light as possible. It is no longer useful to think of clothing as providing greater insulation if it is heavy—it was usual to speak about the fact that someone had a fine winter overcoat because it was 'a good heavy coat'. This is no longer necessary because modern synthetic fibres are low in bulk density compared with fibres such as wool. Similarly, with bed clothes, it used to be thought that more insulation would be provided by increasing the number of blankets. This does not happen in practice and it is better to have a light eiderdown with air trapped between the feathers or a light, so-called cellular blanket. If a heavy blanket is put on top of an eiderdown the overall insulation is almost certainly reduced by compression of the eiderdown and displacement of the air. The modern duvet employes this principle by having lightweight insulation provided, ideally, by down.

Functions of clothing

The functions which are served by clothing include the provision of shelter and Stansfield (personal communication) has tabulated all of these aspects as follows:

Shelter and other individual aspects
Protection—physical (including safety):

(a) From heat/cold.
(b) Mechanical; from abrasion, blows, tears, thorns, etc.
(c) Biological; from animals, plants, germs, etc.
(d) From air/wind.
(e) From water; rain, snow, ice; ponds, rivers, sea, etc.
(f) From fire.
(g) From contamination; dirt, food, oils, grease, noxious chemicals, etc.

Comfort—affected by:

(a) Mechanical pressures and constraints on the body.
(b) Sensory aspects of touch of materials on the skin.
(c) Sensations caused by movement of garment (e.g. swing).
(d) Movement of air around the body through garments.
(e) Movement of moisture around the body and through garments.
(f) Temperatures of the air around different parts of the skin.
(g) Aesthetic; sense of personal elegance.

Effectiveness—aided by:

(a) Providing a base to lie, sit or stand on.
(b) Giving purchase on external objects (e.g. shoes, gloves).
(c) Means of carrying stores, equipment, tools, etc. (e.g. pockets, belts).

Protective clothing

The provision of bulky clothing filled with eider or other down provides superb insulation without great weight and is exemplified by modern polar clothing or the clothing used by mountaineers to go to extreme altitudes. The trapped air may be compressed by external forces or by wind penetrating the garment. It is customary in cold weather clothing to have an outer layer which is described as being wind-proof. This is achieved by having a very closely woven fabric which will still allow moisture and water vapour to pass through. The passage of water vapour is an important characteristic of successful garments; if, for example, a completely impermeable outer layer is worn, such as a plastic oversuit, then the water vapour caused by insensible or sensible evaporation from the skin surface can pass through the clothing but condenses on the impermeable outer layer. If the environmental temperature is sufficiently low, this condensed water will freeze to produce a layer of ice within the plastic oversuit which, depending upon conditions, can increase considerably with further work and condensation. The effect of such ice is to reduce the insulation value of the clothing as the cells become filled with water or ice instead of air. Furthermore, the heat flow through the clothing to this ice layer will be accelerated compared with the situation without ice. It is therefore important to prevent condensation of water within clothing, and it is for this reason that the outer so-called 'wind-proof' layer must be permeable to water vapour. Even so, with the varied rates of working that are common in cold climates, it must be realized that sweating does occur with hard work due to the very effective insulation of polar clothing, which can produce skin temperatures that may be quite high. Condensation within clothing may be prevented by having an adequately permeable outer layer which is reasonably wind-proof. It will be appreciated that the two conditions of water vapour permeability and wind-proofness are in opposition to one another and a compromise is necessary to provide a satisfactory clothing assembly[2, 3, 4].

Compressibility of clothing

The next principle, relating to the axiom that the thermal insulation of clothing is proportional to the thickness of the dead air trapped, is that the clothing thickness has to be maintained by the avoidance of compression. However, it is not always possible to achieve this, especially if the clothing has to be worn while people are standing, walking and then sitting when, inevitably, clothing will be compressed over the gluteal region and probably over the thighs as well. The characteristic to evaluate is the rate of recovery of thickness when compression is removed. In other words, the elasticity of the fabric is important. Although, theoretically cottonwool and fur may be exactly equal to each other in terms of thickness, cottonwool is far less useful because once compressed it remains in this state and has a very slow recovery. Fur, on the other hand, when compressed, rapidly regains its original thickness when the pressure is released. Another factor of importance in clothing in regard to the movement of water vapour is that materials can vary between those that can absorb water such as wool, cotton, etc. and those which do not, such as the majority of plastics.

Considerable advances have been made recently in the development of plastics which are also permeable to water, at present these materials are not altogether successful for clothing because they break easily; no doubt future developments will provide plastics permeable to water whilst at the same time being sufficiently robust for light-weight, cold-weather clothing. Man-made fibres in general may be regarded as impermeable to water, whereas the natural fibres, that is silk, wool, cotton or hair, all absorb water to some extent by a 'wicking' effect. When cotton or wool is dipped into water the moisture will move along the fibre. When most materials are exposed to the environment the important factor is not the absolute humidity of the surroundings but the relative humidity. At an environmental temperature of 0°C, the relative humidity will usually be of the order of 80 per cent but the absolute humidity will be low. Clothing in such conditions will take up water until it is in equilibrium with the relative humidity. When water is taken up by many fibres, there will be a considerable evolution of heat. Conversely, when wet clothing dries there is a loss of heat. Wool has a high heat regain and when a woollen suit or overcoat is worn out of doors in a storm, for a brief period when the clothing becomes soaked the body is protected against loss of insulation due to water saturation of the clothing, by the heat which is given out. This heat regain is less marked in cotton[1].

Layer principle

Overall thickness can be provided by one very thick article of clothing or by several layers. In general, the use of multilayered clothing is better than the single thickness article because the insulation with several layers can be varied easily; a layer can be taken off, or put on, without disturbing the whole clothing assembly. This is useful in cold weather when hard work is being done with consequent sweating; even with well designed clothing there can be condensation within the fabric itself. Furthermore, in cold conditions with hard work and good insulation, body temperatures can actually rise. In these conditions insulation can also be controlled by the use of suitable closures so that the clothing can be quickly opened or closed as required. By opening the clothing, even if one does not take the garment off, the insulation is reduced by changing the effective thickness, enabling any surrounding air movement to affect the clothing itself, and hence the heat loss. The way in which clothing is adjusted is usually by means of zip fasteners of either metal or plastic. In very cold weather, these can freeze or stick. Plastic zip fasteners also have the weakness that in very cold conditions the plastic may become brittle; metal fasteners are probably the most effective. Another form of closure, apart from buttons and toggles, which have their place, is the firm adhesion provided by the material known as Velcro; this has proved of great value in polar clothing and does not have the disadvantage of becoming brittle at low temperatures.

Applications of engineering principles

Some of the ways in which insulation is provided in clothing can be illustrated by examining the dress developed by the Eskimos. This consists of caribou skin, worn as a double layer with the hairs in the outside layers pointing outwards and the hair of the inside layer directed towards the body. This is more effective than the use of hair twice as long as a single caribou hair for a reason which can be found in engineering principles. The bending of a beam under load is proportional to the square of the length. This concept can be applied to the design of clothing by considering hair to be similar to a beam. Caribou hair has another curious characteristic in that the hairs are hollow, and this effectively increases the insulation.

Another application of engineering principles to the design of clothing can be found in the paper published by Stevens[5] (an expert in the design of parachutes) where he described some of the complex physical principles underlying the design of a parachute which are relevant to the design of a skirt to enable it to hang properly. Apart from such rather esoteric applications there are other aspects of clothing where engineering is appropriate. An obvious example is in the strength of fabric and there are a number of standard methods for measuring this. The resistance of fasteners to the strains to which they are exposed is another area where approved technical methods are established. Additional characteristics of textiles, or of complete articles of clothing, can be tested such as their thermal insulation properties, the transmission of water vapour and the ability of fibres to 'wick' water. These methods will not be discussed here as they are described in appropriate technical hand-books[6, 7, 8] (see also Plate 14(b)).

Assessment of clothing

One method to test clothing is to establish the overall insulation value of a complete garment assembly. This is determined by the Clo value which was designed by Gagge, Burton and Bazett[9], so that it could be defined in different numerical systems, and described in simple practical terms. One Clo unit is equivalent to the insulation required to keep a seated subject comfortable at an air temperature of 21°C (70°F) in an air movement of 0.1 m/s. Such insulation is provided by an ordinary suit, with shirt, pants, etc. 1 Clo = 0.18°C/kcal/h/m² (0.155°C.m²/W when defined as mean thermal resistance). It may also be expressed in the unit known as the tog which has the definition 10 togs = 1°C m²/W = 6.45 Clo.

The measurement of the insulation value has been made frequently in large research institutes, and in particular those of the armed forces such as the Institute of Aviation Medicine in England or the Quartermaster Research Laboratories in the USA at Natick. A variety of metal mannikins have been used for this purpose where the surface temperature can be controlled by means of fans and heaters within the model. Sophisticated mannikins can produce differences in temperature distribution between the limbs compared with the trunk, for instance. By dressing these mannikins in appropriate clothing assemblies the effect of insulation on heat flow can be measured directly and very accurately, with the resultant insulation being expressed in Clo units. The use of such mannikins is rather limited because they are expensive to build and quite difficult to dress effectively. They cannot simulate the exact way in which the clothing assembly fits the individual and this markedly affects its insulation value. Since a variety of different fits may have to be tested, this cannot be done with one metal mannikin. Alternative methods have been devised in which the insulation value is determined by direct measurement on the individual. The subject wears a particular clothing assembly and his metabolic heat production is accurately measured, together with assessments of heat flow through the clothing carried out by measuring temperature gradients within the fabric layers together with the use of heat flow discs of the type designed by Hatfield (see Chapter 2). Reasonably accurate measurements can be made in this way of complete assemblies which agree well with the results obtained using metal mannikins. It is probably unnecessary to obtain great precision in the measurement of clothing insulation, since the effect of a garment varies not only with the subject's own heat production, but more particularly with the activity with which the heat production is associated. For instance, if a subject is standing in still air the effective insulation of the clothing will not be the same as if he is in a wind, particularly a buffeting one, when there will be a considerable bellows effect and clothing may be compressed against a particular part of the body and may billow out at another area. Hence the futility of measuring very precise insulation values.

Measurement of trapped air

An aspect of clothing which has been frequently mentioned already is concerned with the volume of air that is trapped, both within the fibres making up the clothing, and also between the layers of clothing itself. To describe the properties of clothing assemblies it is necessary to include a measurement of the total volume of air that may be entrapped as well as the rate of diffusion of such air through the clothing. This can be done by a nitrogen wash-out technique where nitrogen is introduced into the microclimate beneath the clothing assembly whilst it is being worn. The rate of decay of nitrogen concentration as it is lost by diffusion or by bellows action and replaced with air, can be measured with considerable accuracy. Hence the volume exchange rate of the air within the assembly can be calculated. Further details of this technique are given by Crockford[10].

Clothing for cold conditions

Although the overall insulation of a garment may be high the protection of the extremities still provides many problems. The difficulty is to insulate the fingers effectively without reducing their use. This is where the problem arises of insulating small diameter tubes such as fingers, as mentioned in Chapter 3. The lagging of small pipes can sometimes actually decrease insulation when the diameter of the lagged part is significantly increased by the insulating material, thereby leading to a greater surface area available to dissipate heat.

This unfortunate effect means that it is difficult to do anything except enclose the whole hand in a mitt. Effective insulation can, in this way, be provided for the hand but at the cost of dexterity. Various combinations have been used with, for instance, a three finger mitt with a separate space for the first finger and thumb and this can be a useful compromise. It is also useful to wear thin gloves which may not provide insulation but can prevent frost burn. If one touches icy cold metal with a bare hand or finger, the finger can stick to the metal because of immediate freezing of the surface tissue. It is difficult to separate the finger and metal without injury to the skin. The wearing of a thin silk glove can prevent this happening immediately although, of course, with prolonged contact, freezing will occur even when the glove is worn. One can have protection from cold injury in this way even though it is difficult to prevent loss of heat from the fingers and hands and consequent fall of temperature in these parts. Apart from the loss of dexterity, the fingers do not become particularly painful until the skin temperature drops to about 10°C; then discomfort begins and can be quite severe. The same problem occurs with regard to the feet, but since dexterity is not such a serious problem in the feet it is much easier to provide effective insulation by the use of thick boots with proper insoles and layers of socks without much fabric compression.

Protection of head and eyes

The other extremity is the head which presents various difficulties, a specific one being the provision of adequate goggles to prevent cold damage to the surface of the eye itself. With goggles which have been designed so far, the moisture from the surface of the eye condenses on the lens to cause fogging. This is a handicap which has yet to be overcome; there are many different types of goggles, new pairs of so-called revolutionary types being produced at least once every year. However, none so far has proved to be entirely satisfactory in practice. As far as the head is concerned it presents a large surface area and some of the problems of heat flow from the head have been discussed in Chapter 3. The head is generally sheltered by the convective plume of warmed air from the body, but this can be blown away out of doors although it may not be completely destroyed. It is for this reason that thick insulation of the head does not prove necessary in some conditions, and a fairly light covering, as provided by a Balaclava helmet can be quite adequate. In more severe conditions a thick fur cap or fur hat on top of a Balaclava provides very good insulation. No more appears to be required.

The problem regarding cold weather clothing is that there are conditions under which it is impossible to protect the extremities completely, particularly the hands. When dexterity is required there is the alternative of providing electrical heating of gloves. Relatively small amounts of heat are needed to maintain finger temperature and dexterity, and this power can be provided by batteries; the heat required for each hand is of the order of 6 watts. Electrical heating of hands and feet has proved to be a practical way of overcoming severe exposure, especially for people who are working, for instance, on the decks of ships, where they can if necessary plug into an electrical power source and do not have to be hampered by carrying heavy batteries. The provision of adequate clothing for pilots who may be flying at high altitudes with low cabin temperatures presents considerable problems. Fighter pilots have to be protected but dexterity has to be preserved and the bulk of clothing must be kept to a minimum. Under these conditions, electrically heated suits are used and in hot conditions completely air-conditioned suits[11] are also available which use a portable conditioning unit. Air-ventilated suits are also used; air is piped through a series of tubes sewn into the inner surface of the suit, and appropriately distributed to heat the whole body surface and maintain skin temperature at a comfortable level. This means that the temperature in practice will be rather lower along the limbs than over the trunk. This type of clothing has been worn by pilots and somewhat similar garments have been designed for space travellers.

Clothing for harsh environments

One of the harshest environments found in the armed services is that beneath hovering helicopters on ships at sea in arctic conditions. The low temperatures, together with surface winds and the air movement due to the motion of the ship through the sea, coupled with the downdraught produced by the helicopter rotors, create a very unpleasant and stressful environment. Studies of the air patterns produced by helicopter rotors (Chapter 4) have shown that the turbulent air movements cause severe buffeting which leads to extremely high rates of body heat loss, with convective cooling coefficients several times the values found at the same mean wind speeds in linear air flow.

Deck crews who work beneath hovering helicopters to marshal, fuel, arm, service and repair these aircraft must therefore have suitable clothing to protect against this extreme heat loss.

Colour infra-red thermography has been used to evaluate the effectiveness of garments that may be worn in these conditions. Subjects were dressed in a number of alternative garment assemblies and stood on land beneath a hovering helicopter. This study was made in England during the winter at an air temperature of 5°C. Infra-red thermograms were made at various times during this exposure to observe the temperature pattern over the garment surface. Plate 14(a) shows typical thermograms for such conditions; inadequacies in insulation, or of fit, leading to 'hot spots' and excessive heat loss were easily identified. These tests enabled the optimum garment assembly to be selected very quickly. The choice was subsequently confirmed by trials at sea in which clothing and skin temperatures were monitored by thermocouples and where assessments of comfort and acceptability were made from the results of questionnaires.

Plate 14(a) also shows that the face can lose a high percentage of body heat, because of the difficulties previously mentioned, it is an area generally left unprotected. The thermograms show the exact extent of the exposed skin on the face together with its precise temperature. Such information allows calculation of the heat loss from the face as a percentage of the overall body heat exchange. In the conditions found beneath hovering helicopters the face heat loss accounts for up to 40 per cent of the total body heat exchange and in some situations this will be the limiting factor in exposure time.

Protection against cold water

Protection against cold water is important to sailors and yachtsmen who may be shipwrecked or lost at sea (see p. 166). The various ways in which this protection has been provided stems, to a large extent, from the experiences in the Second World War, when many seamen were shipwrecked and spent, in many cases, a great deal of time in the water or in life rafts or life boats before being rescued.

The first important development was that of the inflatable raft. This was sponsored by the Royal Navy and resulted in the construction of rafts with a canopy, which could be packed into a relatively small bag but, on being cast into the sea, would inflate by the release of compressed air or carbon dioxide contained within a cylinder attached to the raft. These inflatable rafts are now familiar but the first model was only tested some 30 years ago. There are now many different patterns but the basic principle is the same in all. It consists of a two layer rubber floor with air in between. There is a gunwale approximately 0.3 m high consisting of two inflated rings and a double thickness rubber canopy covers the whole raft and held in place by inflatable struts. The survivors are thus protected from wind and rain and largely from the waves and spray. The rafts are made in several different sizes to accommodate 5, 10 and even 25 men and are usually oval in shape with apertures at either end which can be opened or closed as required depending on the size of the sea. Provided the full complement of people is aboard the raft, temperatures can rise markedly when both apertures are closed; with an outside temperature as low as $-5°C$ the inside may be in excess of $15°C$. There is thus a most effective thermal protection and men have been able to survive in such rafts under extremely hazardous conditions for several days and have been in reasonably good physical shape at the end of this experience. There are some disadvantages; in very rough seas the floor may buckle, producing an extremely uncomfortable motion which has resulted in trials carried out in very rough water having to be abandoned because the majority of those on board were incapacitated by sea-sickness. However, even though this is a serious disadvantage, in general, when the full complement of men are aboard and are properly instructed in the use of the raft they can often maintain balance and equilibrium and so prevent this motion from becoming intolerable.

Although the provision of inflatable rafts has undoubtedly saved the lives of many people at sea (both fishermen on small trawlers, and those on board much larger vessels), further protection is needed for the man who escapes from a sinking vessel, whatever its size, and who is floating on his own in the sea[12]. He may be provided with a life jacket whose floatation will keep the head above water and so preserve him from immediate drowning, but such protection does not prevent body cooling and it is essential to provide further protection to prevent temperatures falling to dangerous levels. The simplest form of garment has been a plastic coverall, rather like a sack, which is donned before getting into the sea, and by fastening at the neck, will keep the subject, and his clothing, dry and provide good insulation in the sea. Such very simple devices have been tested and undoubtedly protect, but have the disadvantage that they are fragile and can be easily torn or snagged on escaping from a sinking vessel.

Submarine escape
Research into protection for crews who had to escape from submarines unable to surface resulted in the submarine escape immersion suit. The design stems from a disaster in the Thames in the early 1950s when a submarine was sunk in cold weather in relatively shallow water of about 10 m. Many of the crew managed to escape but were carried downstream by the fast-moving river and were not spotted in the darkness, they were only picked up at dawn after some 8–10 hours and most were dead. It seemed probable from post-mortem examination that they had died, not from drowning, although that might have been the terminal cause, but from the effects of hypothermia, due to the immersion in cold water. In consequence, it was decided that some

form of protection was necessary for submarine crews and an immersion suit was designed which consisted of a double layer of impermeable fabric that could be inflated and looked rather like the famous Michelin man. Trials were carried out with seamen wearing such suits floating in the cold water for periods of many hours. In one such trial men floated in water at 5°C for up to 12 hours with falls in body temperature of the order of only 1–1.5°C. With such a suit the subject half floats in the water, usually with his head well up and his arms out of the water with his feet floating up, and tending to assume a 'V' shape with the buttocks being at the lowest point in the water. In mild weather with calm seas, subjects can float and swim on their backs using their arms, at a reasonable pace. In these circumstances they may even become too warm, especially if there is sun. Surface temperatures can vary greatly between the immersed and non-immersed parts of the body; depending on the water temperature there can be differences of the order of 20–25°C.

These protective suits are extremely effective in preventing body temperature drop and preserving people from death at sea. Some of the problems associated with the use of the suit are due to other aspects of cold physiology. Floating in cold water induces a marked cold diuresis—a phenomenon familiar to most people who are suddenly exposed to cold weather. The secretion of urine increases due to the suppression of the anti-diuretic hormone. Subjects wearing immersion suits do, unfortunately, suffer from this cold diuresis and this means that after 2–3 hours they may have a volume of up to 1 litre of urine in their bladders. Not only is this painful, it is dangerous in that there could be rupture of the bladder if it were not emptied. It is essential for survivors wearing these suits to empty their bladders after a period of approximately 2 hours. This has the great disadvantage that it soaks whatever clothing the subject is wearing, which then loses its insulation and can result in very severe chilling. There have been a variety of methods designed to try to alleviate this cold diuresis, the most practicable being the use of 'babies nappies' with rubber pants worn over them. The manoeuvres necessary to don such garments, however, make them extremely difficult to use for someone who has to abandon ship quickly. The immersion suit can be donned rapidly if the 'nappy' is not incorporated, and the suit is provided with a cylinder of carbon dioxide which provides easy inflation. When a cord is pulled the suit blows up and the subject will float.

Hot weather clothing

Clothing is also important in hot climates to protect the individual from radiation, either solar or reflected. In jungles, clothing can also protect against mechanical injury and from biting insects and contact with poisonous plants. In hot climates it is necessary to differentiate between the hot dry desert environment and the hot humid conditions of the tropics, where the temperature may not rise to very great levels, perhaps 36–38°C, but where humidity is high—exceeding 70–80 per cent RH. In such conditions, clothing is a nuisance; it impairs heat loss by evaporation and such evaporation as there is can be of great importance. Convective heat loss will also be affected by clothing as will radiation to some extent, depending upon the particular characteristics of the material itself. In hot humid environments a minimum of clothing is most desirable, and this should be loose and of the lightest possible materials.

In a hot dry environment it is noticeable that people indigenous to these regions, such as Arabs, wear rather bulky clothing which is not necessarily light. Such clothing provides shelter from direct and also from reflected solar radiation. By being fairly voluminous it allows a considerable layer of air between the clothing and the body itself which, coupled with the billowing movement, produces a bellows effect whereby heat is lost in the airstreams which are expelled from the clothing. Those who have tried wearing Arab clothing in hot desert conditions claim that they are quite comfortably cool. The voluminous clothing is also valuable in that temperatures in hot deserts are very high during the day—reaching 45–50°C, although humidity is in general very low. However, over the course of 24 hours with the daylight lasting some 12 hours, temperatures at night fall extremely rapidly, since there is usually a clear sky with consequent high heat loss by radiation. Temperatures drop abruptly in the latter part of the day and they continue to fall after sun-set, and may reach levels which are near to freezing. Under such conditions the Bedouin Arab needs to have considerable clothing, which is useful during rest at night in providing effective insulation. Another feature of the Bedouin clothing is that it also protects against sand storms which can be extremely unpleasant; sand grains may cause considerable pain when driven by a hard wind against the skin and the protection provided by the burnoose is very useful.

One of the more important aspects of hot weather clothing is head gear, which provides protection against solar radiation. It was believed for a long time that there were dangerous so-called 'actinic rays' in sunlight, which had the remarkable power of penetrating through the skin and bones of the skull to cause direct damage to the underlying brain. Therefore, it was considered to be most hazardous to go out in the noon day sun unless one's head was properly protected. This led to the popularity of the sola topi (the word sola refers to the type of cork that was used for insulation in these so-called pith helmets). It was very light in weight and provided effective insulation against strong solar radiation. However, although it is known that there are no such actinic rays; nevertheless, it can still be useful to protect the head, especially for those people who are bald and who are liable to get blisters on the scalp. For those who are not bald, it is also useful to provide some covering for the head, preferably of a white material which will reflect some of the sunlight. It is not that there are dangerous rays but one of the problems in very hot climates is the possibility of heat stroke due to a rise in body temperature.

Industrial protective clothing

There are a number of special situations where specific forms of clothing are required to protect the wearer from injury. These include the problems of intense thermal radiation, as with blast furnaces, or in the manufacture of glass or ceramics. Some protection can be provided by the use of radiation shields, made from polished aluminium, but when it is not practicable to use these the workers need clothing assemblies which insulate them effectively. Such garments may incorporate a layer of aluminium foil to act as a reflector, but they mainly protect by having thick layers of clothing. Wearing a protective garment of this kind in a region with very high temperatures means that the tolerance time may be relatively short, as body temperatures will inevitably rise. The rate of rise will depend on the amount of physical work which is done. The provision of garments ventilated with cold air can increase tolerance times substantially.

The use of well-designed cold weather clothing by workers in refrigeration plants, as in the food industry, is as yet not popular. In a cold store, the work pattern consists of conveying articles into and out of refrigerated spaces. The worker is exposed intermittently to cold, alternating with time spent working outside the low temperature area. The wearing of bulky clothing is hampering and unacceptable in temperate conditions and proper gloves and suitable boots are in general the only specialized items used.

There are many industries with hazards from the inhalation of dust or damaging gases and vapours. In chemical plants, there may be a need for protection from contact with substances which can damage the skin or the eyes. A number of special clothing assemblies have been designed for such conditions and these incorporate various forms of masks or helmets with suitable provision for ventilation. Further descriptions will be found in standard textbooks[13].

The design of different protective clothing assemblies has to be such that heat balance can be maintained. This can be difficult to achieve, as the need for protection and for heat balance can easily be incompatible. The difficulties involved can be illustrated by examples from the armed forces.

It may be necessary to provide clothing to protect against poison gas, specifically the so-called nerve gases. Effective clothing has been designed, which is usually worn as a coverall over the appropriate army uniform. This has to be combined with a gas mask so that the soldier is well protected against nerve gases, but since the anti-gas suit cannot be made permeable to water vapour, but impermeable to gases, there is a serious hazard of collapse due to hyperthermia.

Rather similar difficulties have arisen with workers in contact with radioactive material. At present the solution is essentially an operational one, by doing everything possible to limit the duration of exposure, i.e. the length of time the protective clothing has to be worn.

Clothing for clean environments

The human body is constantly losing skin scales and micro-organisms which are carried away from the skin surface by the natural convection boundary layer flow (see Chapter 4). In hospital operating theatres and special care wards, such as burns units, there is a need to prevent the dispersal of micro-organisms into the air from the skin of the nurses and surgeons. It is now generally accepted that in an operating theatre the main risk for airborne cross-infection to a patient comes from the operating room staff themselves, who constantly shed micro-organisms and skin flakes. There are two main methods of reducing this dispersion and consequent risk of airborne infection; one is by the provision of special air conditioning systems designed to produce clean areas around the patient and to sweep any contamination away. The other method is by the use of special garments designed to be impermeable to organisms and skin scales.

The performance of such garments has to be assessed from two standpoints; in the first place it has to be impermeable to the potentially pathogenic micro-organisms and skin flakes whilst at the same time allowing the passage of air and water vapour so that the wearer does not become uncomfortably hot. The size of weave of an ordinary cotton surgical gown is about 70 μm $\times$ 70 μm and provides little resistance to the passage of skin and organisms. Dispersal of skin scales is enhanced by the 'cheese grater' action produced by the fabric rubbing on the skin surface. In order to restrict the passage of skin and bacteria, materials are sometimes used, made of paper or plastic, which are quite impermeable to these particles. Such garments do not allow the body to lose heat by evaporation and are generally extremely uncomfortable, especially if the wearer has to perform moderate or hard physical work.

The bacteriological and thermal requirements of these garments are to some extent, incompatible and compromises are necessary for workable clothing assemblies. The effectiveness of a garment against bacterial dispersal has been quantified by the use of a dispersal chamber; this was a cabinet some 2 m high and 1.2 m square with a close fitting door at the front. The subject entered the chamber wearing the garments to be tested, and a bacterial slit sampler connected to the chamber drew an aliquot of air over a petri dish containing nutrient agar. Any micro-organisms entrained in the sampled air were deposited on to the nutrient and, when the plate was incubated, visible colonies of bacteria were seen clearly. The subject in the chamber marked time, by walking on the spot for a period of 2 minutes during which the air was sampled at a known rate. The number of bacterial colonies counted on the plate after incubation characterized the dispersal from the garment assembly.

Typical results from such tests can be illustrated with regard to the clothing used within a burns unit[14]. The normal clothing for the nursing staff consisted of a cotton two-piece suit worn for one shift only. As part of the barrier nursing routine designed to minimize the chances of both airborne and contact cross-infection, the nurses wore a cotton or plastic apron over the cotton suit when attending patients. An alternative to the cotton suit was one made of spun-bonded polyethylene which was regarded as semi-disposable and could be laundered approximately 4–5 times before having to be discarded. Figure 10.1(a) shows a subject wearing each of these four garment assemblies and Figure 10.1(b) (top) shows the dispersal of organisms from the body with each assembly as a percentage of the dispersal from the cotton suit alone. The overall comfort and acceptability of the garment assemblies was assessed by asking the nursing staff to 'score' comfort as a percentage of that of the cotton suit, which in this case was assumed to be 100 per cent acceptable; the results obtained are shown in Figure 10.1(b) (lower). The two-piece plastic suit gave the least bacterial dispersal from the body (only some 40 per cent of that from the cotton suit alone) but it was found to have the lowest comfort rating at only 38 per cent of that for the cotton suit.

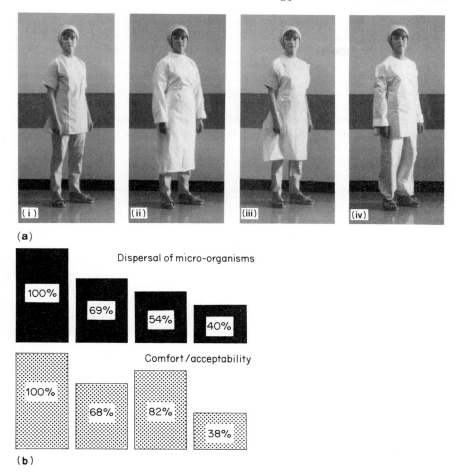

Fig. 10.1 (a) Clothing assemblies suitable for nurses in a burns unit; (i) Cotton two-piece suit; (ii) Cotton suit with cotton gown; (iii) Cotton suit with plastic apron; (iv) 'Plastic' suit made from spun-bonded polyethylene; (b) The four assemblies graded according to bacterial dispersal and comfort. (Reproduced with permission from the *Journal of Hygiene Cambridge*)[14].

Although the plastic suit was comfortable as long as there was little move-ment, the wearer felt extremely hot if there was any appreciable activity; there was no sweat absorption and this added greatly to the discomfort. If a gown or apron was worn over this plastic suit during a burn dressing, for instance, the nurse became almost unbearably hot and when she sat down the suit tended to stick to plastic chair coverings. Another criticism of these suits was that they were poorly styled and the appearance was disliked by both patients and staff. In organisations where clothing or uniforms denote status and authority, this is a particularly important point for the clothing designer to bear in mind.

In the burns unit where these investigations were made the cotton suit worn beneath a disposable plastic apron was found to give an acceptable degree of comfort with a marked reduction in bacterial dispersal compared with the cotton suit alone, and this was adopted as the standard assembly.

Similar investigations were performed to assess the suitability of materials for operating theatre gowns and results are shown in Figure 10.2 which contrast the bacterial dispersal from subjects wearing cotton, paper and plastic gowns over cotton suits. One way of overcoming the thermal problem associated with impermeable gown materials is to provide mechanical ventilation at the skin surface. This principle is used in the Charnley-Howorth Body Exhaust Gown (Figure 10.3). This consists of a very close weave, all enveloping, one-piece gown which is almost impermeable to organisms and skin scales. The surgeon wears a special head-piece and visor connected to a suction pump; an airflow passes over the whole surface of the body to take away heat, moisture and potentially pathogenic particles. The effectiveness of this garment at reducing bacterial dispersion compared with more conventional gowns is also seen in Figure 10.2. Such a gown represents a radical departure from conventional operating theatre clothing and requires special air extraction facilities which can easily be incorporated into new ultra-clean air operating facilities, or provided by portable equipment for general hospital use.

Other experiments made in the bacterial dispersal chamber indicated that many of the potentially pathogenic organisms liberated from the body originated from the perineal region. Dispersal of these organisms was greatly reduced if the subject wore underpants made of a close weave cotton (Ventile) material similar to that used for the body exhaust gown.

In addition to the physical properties of the material from which operating room garments are made, the 'fit' and methods for making closures also have a bearing on bacterial dispersal and thermal comfort. The 'bellows' action of garments during movement enables air exchange to take place within the clothing layers. Schlieren photography has been used to visualize this 'pumping' action which can blow organisms directly into the operating theatre atmosphere from the microclimate near to the skin surface. This effect occurs at most non-occluded openings (the bottom edge of the gown, trousers and at openings around the cuffs and neck) as shown in Figure 10.4, and can be overcome by occluding these openings whenever possible, for instance by the use of elastic collars and cuffs.

Disposable clothing

For many applications the use of disposable or semi-disposable garments appears to be an attractive proposition. However, there are a number of disadvantages to the use of such garments. For instance, with garment assemblies specifically designed to protect against nerve gas the staining of the garment by oil will reduce the permeability of the fabric considerably rendering it unsuitable to provide protection against the gas.

In hospitals, the use of disposable gowns and suits is usually prohibitively expensive. The choice is not simply one of laundering conventional cotton

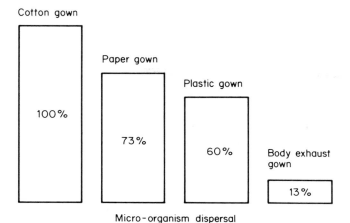

Micro-organism dispersal

Fig. 10.2   Bacterial dispersal from four types of surgical gown when worn over a cotton two-piece suit.

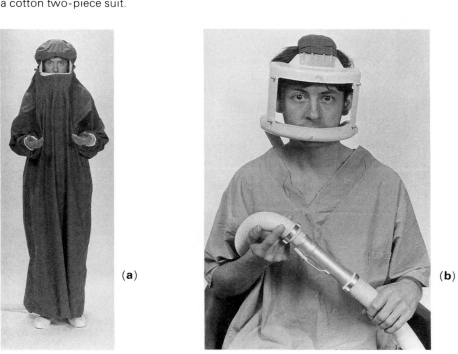

(a)

(b)

Fig. 10.3   The Charnley-Howorth Body Exhaust Gown. This consists of an all-enveloping one-piece gown (a) made of close-weave material. Beneath the gown, the surgeon wears a special headpiece and visor (b) connected to a vacuum air system which produces an upward airflow which removes heat, moisture and potentially pathogenic particles.

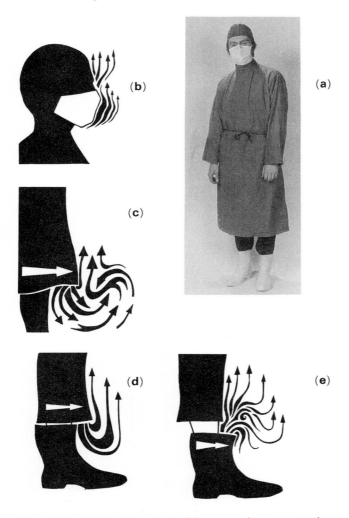

Fig. 10.4 (a) Traditional clothing worn by an operating team during surgery consisting of a cotton gown over a cotton two-piece suit together with hat, mask, gloves and boots. Diagrams derived from Schlieren observations illustrate ways in which pathogenic particles may be shed from the body to the surroundings. (b) A surgical mask reduces the force of expired air and diverts much of the flow upwards to join the convective boundary layer flow. (c) The bellows effect at the bottom of a surgical gown can vigorously displace air from beneath the lower edge to mix with the convective boundary layer flow and room air. (d) Air may also be forced out of the trouser ends and from a loosely fitting surgical boot (e) by movements of the leg.

garments against purchasing disposable garments; even if many disposables are used the need for a laundry within a hospital is always likely to remain. The unit cost of laundering garments using existing faciltities is generally extremely low and often two orders of magnitude lower than providing disposable garments.

The use of semi-disposable garments, that is garments which can be laundered 4–5 times before becoming unserviceable, also meets with difficulties in practice. Such garments are rarely as attractive as when new even after one wash; this generally means that after 2–3 washes the wearer will consider them aesthetically unacceptable and will dispose of them prematurely.

References

1. Burton, A. C. and Edholm, O. G. (1955). The thermal insulation of the air. In: *Man in a Cold Environment* pp 47–72. Edward Arnold, London.
2. Woodcock, A. H. and Dee, T. E. (1950). *Effect of moisture on transfer of heat through insulating materials*. Report No. 170. Office of the Quartermaster General, Environmental Protection Section, Washington.
3. Woodcock, A. H. (1964). Clothing and Climate, Chap. 21. In: *Medical Climatology*. Elizabeth Licht, New Haven, USA.
4. Adam, J. M. (1967). Climate and clothing. *The Practitioner* **198,** 645–50.
5. Stevens, W. H. (1953). Scientific aspects in the design of a full skirt. *Journal of the Textile Institute* **44,** 247–57.
6. Renbourne, E. T. (1972). *Materials and Clothing in Health and Disease*. Lewis, London.
7. Newburgh, L. H. (1949) *Physiology of Heat Regulation and the Science of Clothing*. Saunders, Philadelphia.
8. Freeman, M. T. (1962). *Protective Clothing and Devices*. United Trade Press, London.
9. Gagge, A. P., Burton, A. C. and Bazett, H. C. (1941). A practical system of units for the description of the heat exchange of man with his environment. *Science* **94,** 428–30.
10. Crockford, G. W., Crowder, M. and Prestidge, S. P. (1972). A trace gas technique for measuring clothing microclimate air exchange rates. *British Journal of Industrial Medicine* **29,** 378–86.
11. Bewley, A. D. and Short, B. C. (1976). Liquid conditioned system for aircraft. *Symposium on heat transfer and the human body. March, 1976*. The Institute of Mechanical Engineers, London.
12. Crockford, G. W. (1973). Protective clothing and equipment. In: *Occupational Health Practice* pp 360–78. Ed by R. S. F. Schilling, Butterworths, London.
13. US Handbook of Public Health.
14. Clark, R. P. and Mullan, B. J. (1976). Clothing for use in clean environments. *Journal of Hygiene Cambridge* **77,** 267–69.

11

The Built Environment

'I think the real function of a building is to keep my papers from being blown about.' J. D. Bernal[1]

Design principles

The construction of a building generally requires the services of an architect, part of whose responsibility are the plans for the provision of lighting, heating, ventilation and the control of noise.

McIntyre[2] has emphasized that 'decisions about the thermal environment in a building must be made at the design stage ... it is no longer acceptable for the architect to call in the heating engineer only at the last moment'. He states that 'the most important factor defining comfort [in buildings] is warmth'. There will be different solutions to the problems of constructing a factory, a school, a block of flats, offices or individual houses, but, in all cases, the building can be regarded as a climatic modifier, providing shelter from wind and rain, diminishing climatic swings and reducing noise.

Specifications concerning temperature, air movement, ventilation, lighting, sound levels and space required per individual are included, *inter alia* in *The Architectural Journal*'s metric handbook, but in many buildings, serving different purposes, indoor conditions are not satisfactory. In some cases this may be due to the use of old buildings with inadequate systems for controlling indoor conditions. Poor construction, deterioration within modern buildings or the need to economize on fuel may also be responsible for diminished comfort. The complexity of the situation can be illustrated by considering schools, where there is an increasing need to reduce running costs. Schools in the UK range from Victorian buildings to a variety of contemporary structures.

The use made of these buildings depends on the teaching which can vary from day to day and between schools. Pupils frequently change classrooms, and each class differs in numbers, in age and proportion of boys and girls. The school hours are usually 09.00–16.00 but the premises may be used in the evening. There are holidays, when adult classes may use the building. It is difficult to provide temperature criteria for such a complex situation and to decide where temperatures should be measured and where thermostats are ideally placed.

Auliciems[3] measured temperatures in a number of schools and examined the relationships between the pupils behaviour and indoor and outdoor conditions. Indoor temperatures varied widely but there were few complaints, probably because the comfort zone for children is distinctly lower than for adults, due to their higher resting metabolism. Children seldom stay still, even when sitting, and this activity increases their heat production.

A minimum temperature of 18°C should be prescribed and the heating system designed so this temperature is reached by the time the school opens. O'Sullivan and Cole[4] studied the thermal performance of two well-designed primary schools over a nine-month period, during which satisfactory conditions were maintained. When the outdoor temperature was 0°C it was 20°C inside the school. There were diurnal temperature swings between 09.00 and 15.30, when the buildings were used. In February the swings were 19.2–21°C in one school and 18.5–20.5°C in the other. In July at the perimeter of the buildings the swings were 21.2–27.8°C and 21.0–25.9°C.

The main criticism was that the heating system had been designed to produce too much energy. O'Sullivan and Cole point out that the thermal energy patterns and the time lag in the fabric makes it almost impossible to calculate the heat balance.

Solar load

Amongst the factors which make calculation of heat balance difficult is the variability of solar radiation. Longmore and Ne'eman[5] and Ne'eman[6] have analysed sunshine incidence over a ten-year period in London and have studied preferences for sunshine. Their results are set out in Table 11.1. There are widely differing views but in general sunlight is liked if it is not associated with thermal or visual discomfort.

Table 11.1 Preferences for sunlight—note the marked contrast between staff and patients in hospital. (Adapted with permission from Longmore and Ne'eman[5].)

Subjects	Preference for sunlight	
	'Like' (%)	'Dislike' (%)
Housewives	93	4
Office workers	73	24
School pupils	42	52
Hospital patients	91	2
Hospital staff	31	62

Interaction between buildings and the surroundings

Chandler[7] describes and discusses 'the man-modified climate of towns'. Buildings in urban centres inevitably change the climate outside as well as inside the buildings. 'The effects of the urban surface's complex geometry, the shape and orientation of individual buildings, the thermal and hydrological properties of buildings, road and other urban fabrics, the heat from metabolism and various combustion processes operating in the city and pollution which changes the chemistry of the urban air, all combine to create a climate quite distinct from that of extra-urban areas.' Air movement in cities is extremely turbulent due to the irregularity of the surface. There is a complex interaction between the effects of the elevation and alignment of different buildings on air movement, producing a unique pattern for each city. Some generalizations are possible; wind speed at the 'surface' is less than in rural areas but may increase sharply with height. In London, surface wind speed averages 6–13 per cent less than in adjacent rural areas.

There is also a temperature build-up over urban areas leading to the formation of 'heat islands'. These have been photographed using the technique of infra-red thermography and some striking pictures of these 'heat islands' have been published in the *National Geographic Magazine* for March 1981. The warm air build-up, as Chandler shows, is greatest at night when after sun-set there may be urban–rural differences of 5–11°C. City size is not so important in this heat build-up as the density of the building development.

Tropical buildings

The design of a building for tropical environments has been discussed by Marcos-Assaad.[8] He defines tropical environments as areas where heat is the dominant problem and where buildings serve to keep occupants cool rather than warm. He stresses that urbanization in the tropics cannot be solved by Western technology, because of economic reasons. Nevertheless, today tropical buildings are being erected which could perfectly well be found in Glasgow or Chicago. Marcos-Assaad points out that the resulting environment is climatically and socially unsatisfactory. One of the problems is that the climate affects buildings which in turn react on the climate. In considering the whole question of the built environment, it is important to realize that thermal comfort is not necessarily the same as thermal equilibrium.

In hot conditions the objective of building design is to prevent heat gain and to maximize heat loss with the removal of excess heat by mechanical cooling. Where conditions vary between hot and cold, the function of the building is to smooth out the diurnal variations both to prevent heat loss in cold periods and to minimize solar gain in the hot period. This requires a flexible mechanical heating and cooling system.

In general, the balance equation for a building is similar to that of man. The differences are that man is not supplied with openings through which air may pass in either direction nor does he have an inflexible system; his heat production can be varied from minute to minute and heat loss can be varied according to the demands, both of his own heat production and the environmentally imposed conditions.

Specifically, this means that the heat flow in human tissues is, nearly always outwards, from the deep layers of the body to the skin surface and hence to the environment. In a building the heat flow in the surface layers, or in the fabric itself is variable and can change over short periods of time. With intense insolation and occasional clouds the heat transferred through or to the surface of the building can be high at one moment and be reduced almost to zero within a few minutes. Inevitably heat fluxes are established within the fabric in both directions which are difficult to calculate and to measure. Hence the accurate heat balance of a building is much harder to assess than the heat balance of a human being. Nevertheless, fairly simple calculations can give the essential level of information about the building's heat balance and in many respects the equations are the same as those which may be used for man.

New techniques have been applied to this problem, in particular the use of infra-red thermography to identify areas of excess heat loss or production, which can be mapped just as they can over the body surface of man.

Some of the ways in which tropical buildings can be used to make the inhabitants more comfortable are described by Marcos-Assaad particularly in terms of traditional building methods. Having dismissed the use of Western architectural practices he points out that the traditonal methods are effective in using the temperature differences inherent in hot countries between day and night. The techniques include the use of curved surfaces to increase radiation exchange; the use of louvres and venetian blinds and awnings to utilize whatever air movement there may be to ventilate the building. Vegetation is used to provide both shade and to act as windbreaks against fierce desert winds. Roofs can be designed as sleeping areas to be used in the first part of the night until temperatures have dropped sufficiently for the indoor climate to become preferable. Air scoops may be used to bring air into a building, and subsequent cooling by passing it through or over water can be most effective as ventilation when combined with high chimneys for removing hot air into the upper-atmosphere (the stack effect).

Using these methods, satisfactory indoor conditions have been achieved without the use of expensive, power-consuming air-conditioning equipment. The environment provides all the power necessary to effect adequate cooling. Warm, humid equatorial regions with temperatures ranging from 21 to 32°C with little seasonal change, produce discomfort associated mainly with a wet skin, that is with unevaporated sweat. The variant of this climate, known as Island Equatorial, can provide steady trade winds, which can be used for effective ventilation in buildings. In hot dry desert regions there are large diurnal and seasonal changes, and here the use of landscaping with courtyards, fountains, etc. produces a very effective control of the hot environment.

A different climate again is the monsoon where there may be three or four seasons of the year varying from hot dry to wet humid. Here it is difficult to design buildings which are satisfactory at all times.

Philosophical vs practical design considerations

The views of the architect in producing a satisfactory indoor environment have been set out by Geoffrey Broadbent in his book entitled *Design in Architecture*[10]. This contains some of the relevant psychological attitudes but the physiological side, although not ignored, is given less weight than a professional physiologist would require.

Broadbent quotes Le Corbusier as pointing out 'that all men have the same organism, the same functions and the same needs' which overlooks the awkward fact of individual variation. Man shows marked variation in body size, weight and contours and hence in movements which affect his use and appreciation of any building. Moreover, there are differences between men and women, there are the effects of age and the differences between the infant, the child, the adult and the elderly, as well as individual variations in body temperature, chemistry and metabolism. It is worth emphasizing the fact of individual variability in view of Le Corbusier's adage 'a house is a machine for living in'. This implies a standard man with standard size and reactions: such a person does not exist.

Le Corbusier has also stated that '. . . by needs, I mean utility, comfort, and practical arrangements . . .' and he was quite clear that needs could be established by observation and by statistical analysis. Such an attitude must imply the similarity of individuals rather than the differences that we are emphasizing. The architect is anxious to have standards established and Gropius is quoted by Broadbent as writing that, 'a Standard may be defined as that simple practical exemplar of anything in general use which embodies a fusion of the best of its anterior forms. Such an impersonal standard is called a norm—a word derived from a carpenters square.'

A list of factors on which architecture can be based was drawn up by Hanneus Meyer, the director of the Bauhaus. His list includes sex life, sleeping habits, pets, gardening, etc. There is no mention of physiology. The arrogance of considering, as Meyer did, that these are the only requirements that need to be considered when building a house, perhaps makes it not surprising that there have been so many disastrous buildings in recent years, at any rate from the point of view of the physiologist. Broadbent himself strongly criticizes this attitude of Meyer, and says 'our tasks as designers of architecture are to reconcile the needs of certain human activities with a given climate by controlling that climate through the medium of the building. The building will be real, physical and three-dimensional; it will enclose spaces which are appropriate in terms of space standards, environmental control and so on for the activities which are to take place there. It will also contain various means by which the user can move or otherwise communicate from space to space.'

This statement implies that the essential function of a building is to modify the external climate to provide a satisfactory indoor climate. The difficulty is to allow for individual variation and the pattern of physiological responses which are involved in the body temperature regulating mechanisms. The architect stresses in many cases, the psychological problems; the feeling about space, the need or otherwise for daylight and the general sensations which are inevitably subjective.

The physiological aspects are described by Broadbent as 'the human being is acting as a multi-modal sensing organism. All the senses are stimulated simultaneously; each perceptual act is a transaction between a stimulus and experience.' These features have been emphasized in discussing the various components of the thermal environment that can be combined to give an index of the thermal state. The components can be measured individually but the human being senses them not separately, but as one integrated stimulus. The other senses of sight, sound and smell can modify our appreciation of body temperature. If one is trying to assess the thermal state of an individual in a building, it is important to measure all the factors which act on our sensory system. Nevertheless, this can be exaggerated because in a temperate climate, the air temperature itself accounts for virtually all of our thermal sensations. These points have been dealt with in Chapter 8 where thermal comfort indices have been described.

These facts imply considerable difficulties when incorporating the results of experiments on people into environmental design. This is largely because most results are obtained by the use of questionnaires which are in many ways unsatisfactory for assessing human responses. To avoid the pitfalls, a questionnaire has to be tested many times to be certain that its form is sufficiently neutral, that the questions can be administered verbally without influencing the reply and to ensure that the replies are consistent with the same subject on different occasions.

Another difficult factor in reconciling human physiology and architectural design is that human beings adapt to changes in the environment. This has been described in some detail in the section dealing with acclimatization to heat and, to a lesser extent, cold. There is also adaptation to changes in noise and lighting levels. It may fairly be said that the human being possesses a flexible physiological system which maintains a steady response in spite of a variable environment. By adaptation, the human has an extra dimension of control.

This conflict between physiological and architectural attitudes is exemplified by the fact that, according to Broadbent, 'people will bring their attitudes or prejudices to the building. Their past experience is an essential part of the perceptual translation.' It is for these reasons that it has proved so difficult to lay down reasonable standards for comfort in buildings.

Human sensitivity to weather

The effect on and susceptibility of man to various forms of weather have been the subject of speculation for many years. There is still little information available which can be regarded as useful or reliable. All over the world, wind is credited with some remarkable effects. Many regions have a particular named wind and it will be claimed that when it blows behaviour patterns are markedly changed. Examples of such winds are the monsoon blowing over India, the Khamsin in the Middle East, the Sirocco and the Föhn in the Mediterranean and Switzerland, and the East wind in England. In Canada and the USA there are the Chinnock and Santa Barbara winds. To take an example, the Khamsin wind is considered by many who live in Israel as responsible for influencing the suicide rate, increases in road accidents and many surgeons are believed to refuse to operate when this wind is blowing. The Sirocco and the Föhn winds are credited with marked changes in suicide rates, bankruptcies and on judgement generally. One suggested reason for these effects on human behaviour is that they are associated with changes in the humidity of the air. In general these are drying winds, there is a fall in the humidity which in turn may affect the ionization of the air. The effects of ionization on man have also been the subject of much discussion and a good deal of research. Many claims have been made for the benefits conferred by negative ionization; for example it is stated that there are marked improvements in perception and in the state of well-being and that negative ions affect the respiratory tract and increase resistance to infection. Recently, there has been renewed interest in the effects of positive and negative ionization as exemplified by two letters in *Science* where the authors, Bissell *et al.*[11], claim that air ions are able to affect a wide variety of living forms and it would therefore be surprising if they were not able to act on humans. Work in Israel is cited as showing that during the Khamsin there are marked changes in, for example, serotonin metabolism. However in other laboratories such findings have not been confirmed: in spite of the considerable claims made for the importance of air ions such effects must be regarded as not proven at present.

This problem and the effects of winds do not exhaust questions about weather. It is commonly claimed that people suffering from rheumatism are susceptible to changes in climatic conditions. Such claims have proved so far to be hard to substantiate. It is difficult to see by what means such weather changes can be appreciated by man. We have no sensors for humidity; all we can sense is that in hot humid conditions the skin may be covered by a film of unevaporated sweat. Most people dislike having such unevaporated sweat and this dislike may enable them to tell differences in humidity. However, it seems probable that there must be some adaptation or acclimatization to such conditions as so many millions of people live in hot and very humid conditions such as the Ganges Valley in India, Bangladesh, West Bengal or many areas in Indonesia. We need to know more about the way in which people may or may not adapt to the discomfort of unevaporated sweat on the skin. Adaptation may consist of ignoring such unpleasant sensations. Nevertheless, we still have the problem that there are many individuals who claim that they can identify weather changes in advance.

Such observations as have been made on weather sensitive individuals have produced equivocal results. There have been suggestions that combinations of changes may be associated with discomfort particularly in those suffering from rheumatism. The term 'rheumatism' is commonly used in the widest non-specific sense to include cases of rheumatoid- as well as osteo-arthritis and other forms of arthritic disorders. These conditions are different from each other and it does seem unlikely that weather sensitivity might be shared amongst them all.

Dampness and humidity

An important aspect of the built environment is the problem of humidity. Words such as damp and chill convey very clear impressions to most people and they are qualities which they prefer to avoid indoors. One of the most serious problems in building today is derived from the use of concrete which has a poor moisture permeability and, in addition, is not a good insulator; heat is conducted fairly rapidly through concrete compared to brick and condensation can quite easily occur on the inside walls of concrete constructions. Buildings with high condensation rates appear to be unacceptable as human dwellings. The main objection to condensation is discomfort. Moist or damp walls may be cold owing to evaporation of water and the discomfort could be due to increased radiation from the body to the wet wall. It seems unlikely that the humidity of the air can be responsible.

Clothing will absorb moisture until it comes into equilibrium with the water at the appropriate relative humidity—it is the relative rather than the absolute humidity to which the clothing responds. This water uptake by clothing can reduce its insulation and hence give, perhaps, some feeling of cold. Here again it seems unlikely that this is the whole story. It is more probable that the objection to condensation is the resulting damage to wall-papers, to painted surfaces on walls and ceilings and to the destruction of curtain materials and floor coverings which become sodden. It is more the unpleasantness from this point of view rather than thermal discomfort which it is suggested produces the dislike of a damp, or even a wet room. It has always been considered, in the popular view, that in conditions of condensation people will suffer from rheumatic fever, tuberculosis and other diseases. Again, the evidence does not seem to be very good. The question of damp is similar in that it involves water in the walls. So-called rising damp usually comes from the fact that the foundations do not have a damp course and so moisture from the ground can seep upwards in the brickwork, or even in concrete, and cause damp inside the building. Such rising damp does not involve as much water as condensation but again the same claims are made that this is extremely uncomfortable and provides an unhealthy atmosphere. It is not intended to argue that damp walls are a 'good thing' from a health point of view but the evidence that it is a 'bad thing' is somewhat scanty.

Some causes for discomfort in buildings

Any heating system in which the air temperature near to the head is greater than that near to the feet at the floor is undesirable and such environmental conditions tend to produce feelings of 'stuffiness' in the head whilst the feet are cold. As long ago as 1857 this was recognized and discussed in a report to the General Board of Health by commissioners appointed to enquire into the warming and ventilation of dwellings[12]. At that time there seemed to be a general understanding of the conditions that were necessary in order to have a comfortable and healthy environment. These, quoted from the 1857 report, are as follows:

1. That the floor be at the highest temperature in the room.
2. That the walls be higher or as high in temperature as the general temperature in the room.
3. That the general body of the air in the room be at a genial and equable temperature at the same level, and gradually decreasing from below upwards to the ceiling.
4. That the ceiling be a temperature differing very little from the stratum of air immediately below and should be the lowest in the room.
5. The range of temperature during the 24 hours be small so as to prevent the loss of heat during the first part of the day by heating air at a much lower temperature than is required.

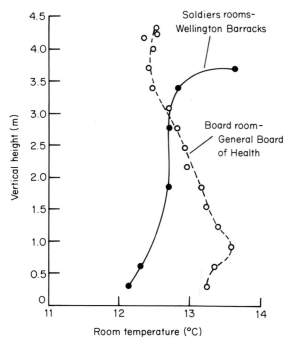

Fig. 11.1   Vertical temperature gradients plotted from measurements made in the soldiers rooms at the Wellington Barracks and the board room of the General Board of Health in London in 1857[12].

6. That the water present in the air be such in amount that no undue checking or accelerating evaporation from the skin takes place.
7. That all heated air which passes upwards should pass away.
8. That fresh air be admitted to supply the place of the exhausted air.
9. That there should be a freedom from sensible currents of air, and a freedom from smoke.

In 1936, Bedford[13] noted that little attention seemed to have been paid to the recommendation of avoiding excessive temperature gradients in buildings, and that systems of heating which created considerable temperature gradients causing warm heads and cold feet had been widely used. Bedford's remarks could be reiterated today when many of the systems for heating and ventilating in buildings produce conditions often quite opposite to those suggested, so many years ago, as conducive to comfort.

The 1857 enquiry included details of the vertical temperature gradients in the Wellington Barracks in London and the board room at the General Board of Health. Figure 11.1 shows that the barracks had the warmest air nearest to the ceiling, but in the board room the air nearest to the ceiling was the coolest in the room.

Without adequate ventilation near to the ceiling, rising hot air from heating appliances 'and people below cannot escape and so builds up to give the greatest air temperature at the highest point in the room. These conditions with excessive vertical temperature gradients often produce complaints of discomfort particularly in rooms where the gradients exceed 3°C. Feelings of cold discomfort are often centred around the limbs, an explanation of this may be given in terms of the gradients and their relation to the human convective microclimate. For a standing man, the natural convection boundary layer flow is thin at the feet and lower limbs and much thicker over the face and head. The thinner the boundary layer, the steeper the temperature gradients between the skin and air and the greater the heat loss. Proportionately more heat is lost from regions where the boundary layer is thin (i.e. the lowest part of the body) compared to the face and head which is well insulated by the wider thermal up-currents. In a room with a vertical temperature gradient with the coolest air near the floor, this effect would be enhanced by increasing the tempeature difference to the skin or clothing to give a greater heat loss. Over the upper part of the body, warmer ambient air produces shallower gradients, and less heat loss. The net result is to increase the difference in heat exchange between the lower limbs and the upper parts of the body. Besides producing an actual increase in heat loss near the limbs the 'natural' differences between the heat exchange with the surroundings for the top and bottom parts of the body are enhanced and these two factors together can lead to overall judgements of discomfort.

Ventilation systems

In many modern buildings heating and ventilating are combined with air input and extract systems. The air ducts are often located in the ceiling and input rates of $8.5\,m^3$/min in rooms of around $71\,m^3$ are quite common. In such systems, extract ducts take away this same quantity of air (possibly a little more if the room is required to be at slight negative pressure with regard to the corridors), the resulting air movement inevitably produces higher convective cooling coefficients for the human body than if the air were 'still'. In addition, high level entry of warmed and conditioned air generally means that the warmest air is high up thereby producing, from the outset, high vertical temperature gradients, which combined with the air movement result in discomfort.

Lecture theatres and classrooms ventilated in this way are subject to a further complication when the occupancy from one period to another may vary from full to practically empty. Ventilation air exchange rates suitable for the thermal load when the room is full are far in excess of that required when it is occupied by only a few people. Then such areas may be excessively uncomfortable due to the increased heat transfer from the body caused by the fast moving airstreams.

An example of the way in which such problems may arise in air-conditioned areas may be seen from calculations involved in producing a room suitable for carrying out infra-red thermography. A room some 7 m square and 3 m high in a bungalow building with windows along one wall facing east was required to have a range of temperatures between 18 and 28°C (± 1°C) at any time of the year in the UK. Air-conditioning calculations for these conditions are generally based on the extremes in the outdoor environment that are likely to be encountered in order that the ventilation equipment specified may have adequate thermal capacity. In this particular room (accounting for a flat roof, window area, aspect, etc.) maximum heat gains of 44 000 British thermal units (BTU) per hour (12.9 kW) were estimated and of this some 25 300 BTU per hour (7.4 kW) were due to the direct solar load.

The plant that was specified as capable of conditioning this area was to consist of two split air-conditioning units each with a cooling capacity of 24 000 BTU per hour (7.0 kW) and a heating capacity of 17 900 BTU per hour (5.25 kW).

The air movement rates in these units were designed to recirculate between 37 and 38.2 m³/min which was some 28 per cent of the total room volume/minute, equivalent to approximately 17 air changes per hour. Such a high air movement rate would produce localized air currents at fairly high air speed to cause draughts and discomfort and in particular would modify exposed skin temperatures and be unsuitable for thermography. The compromise solution was to reduce the air-handling capacity by half (cutting the air change rates by half) and to baffle the air movements through the heat exchangers to produce as even a distribution of air as possible. This compromise meant that the lowest air temperatures required would not be available on some of the hottest days.

In some ventilated or air-conditioned areas a large fresh air supply is needed, as in laboratories and hospital wards where noxious gases or fumes may accumulate or smells may be a nuisance. In other areas 'fresh air' need only be supplied to ensure that CO_2 does not rise above 0.5 per cent and that socially undesirable smells are effectively removed. Experiments carried out on board submarines in the Second World War showed that surprisingly small air change rates would adequately control antisocial smells. Exchange rates of one change per hour produced conditions which from time to time could be unpleasant, but when this was increased to two changes per hour such problems were virtually eliminated.

Two specific examples of the built environment where physical and physiological interaction is important are in hospitals and in chemical and biological laboratories. The degree of sophistication in the control of environmental conditions within hospitals varies widely. Some may be fully ventilated throughout with sealed windows (although full air-conditioning is rare in the UK) whereas others simply rely on natural ventilation through opening windows. An example of high technology applied to the hospital situation is 'The Mount Vernon Project' (page 230). A study of this project and others has suggested that the capital and revenue costs of complex systems are hardly ever justified for the majority of hospital areas. However, there is a critical need to control conditions in operating theatres, intensive care wards and isolation units. It is in areas such as these that isolation or segregation from the rest of a hospital is essential.

Hospital operating theatres

In an operating theatre, cleanliness is vital to prevent cross-infection. Surgical and nursing disciplines are carefully prescribed, rehearsed and practiced to avoid contamination by contact. Of equal importance for the transmission of infection is the air route, and airborne contaminants, from the general surrounding air or from the surgical and nursing teams, can pose a real threat to the successful outcome of a surgical operation. To minimize the risks of airborne organisms, operating theatres are usually supplied with filtered air under positive pressure, to prevent the ingress of dirty air from the surrounding areas. The design criterion generally regarded as being satisfactory[14] is that the quantity of air should maintain this positive pressure when one pair of double doors are wide open, for instance when a patient on a trolley is wheeled into the theatre; an averaged sized operating theatre requires between 43.0 and 56.5 m³/min of filtered air.

When it is necessary to humidify the input air, it is most important that the type of humidifier used does not have reservoirs of water which are excellent breeding grounds for bacteria. There have been incidents of cross-infection caused by micro-organisms spread throughout the air by water humidifiers and Legionnaire's disease is considered to be spread by this route.

Some operating theatres have 'ultra-clean air'. This sophisticated system of air filtration has been used for hip joint replacements, and open-heart surgery where patients may be on immunosuppressive chemotherapy. In such surgical procedures the risks of airborne and contact infection can be great and the air filtration systems need to be complemented by isolation of the surgical and nursing team by special impermeable garments. It has now been shown that the very low airborne concentrations of micro-organisms in ultra-clean air systems can lead to a reduction in sepsis rates[15].

In providing suitable heating and ventilation in operating theatres the requirements of bacterial cleanliness and thermal comfort must be considered together and not separately. In the past, designs have been suitable from one or other standpoint only and thermal comfort considerations have often been incompatible with those of air hygiene, particularly in terms of air movement rates.[32]

Studies have been carried out to assess the preferred temperatures in operating theatres and Figure 11.2 shows some results. In the operating room it is usual for the surgeon to determine the environmental temperature but the Figure shows the preferred temperatures for both surgeons and anaesthetists. The values shown apply for 50 per cent relative humidity with a turbulently ventilated operating room having a mean air movement of 0.13 m/s (25 feet per minute). The surgeons stand beneath the operating lamps and are exposed to more direct radiation than the anaesthetists. It is estimated that the excess temperature due to the operating lamps is 1.4°C. When this study was repeated in the absence of such radiation from the operating lamps the temperatures for the surgeon were raised by approximately 0.5°C. These figures indicate a difference in comfort temperatures of 5°C between surgeons and anaesthetists; the anaesthetists probably prefer the higher temperatures because of the more sedentary nature of their work.

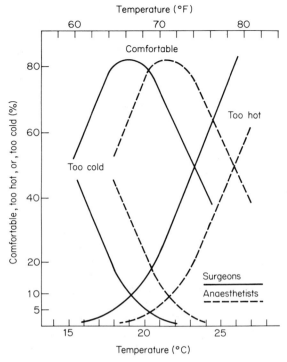

Fig. 11.2 Percentage of individuals comfortable, too hot or too cold at the given ambient air temperatures. The full lines refer to the surgeons and the broken lines to anaesthetists. The values apply at 50 per cent relative humidity with 0.13 m/s air movement. The figures for surgeons include the effect of excess radiation temperature, due to the operating lamp, of 1.4°C. In the absence of any such excess radiation the temperatures for surgeons will be raised by approximately 0.5°C.[14]

Surgical hypothermia

Morris and Wilkey[16] and Morris[17] reported studies on the temperatures necessary in operating rooms to avoid surgical hypothermia. Lightly anaesthetized adults in environmental temperatures of 21–24°C were able to maintain oesophageal temperatures above 36°C during surgery. When room temperature was between 18 and 21°C the oesophageal temperature of all patients studied fell below 36°C after 2 hours of anaesthesia and operation. They concluded that rooms should have temperatures in the range 21–24°C.

The Mount Vernon Project

In 1974, at Mount Vernon Hospital, near London, a three-year clinical trial was undertaken to assess the use of special air support systems in the treatment of major burn injury. A burns unit was constructed consisting of a fully air-conditioned building with 12 beds, and was equipped with sophisticated air ventilation systems designed to provide an acceptable thermal environment both to badly injured patients and the staff who attended them. The air systems were required to produce a bacteriologically clean environment to help minimize airborne cross-infection.

The unit had four special care rooms where patients were nursed during the initial period of their treatment. Suspended from the ceilings were canopies having fabric diffuser tubes which provided a linear down-flow of 'ultra-clean' air at a speed of approximately 0.55–0.65 m/s. The term 'ultra-clean' denoted air with an average of less than 0.1 bacteria per 0.3 m³ some ten times cleaner than air supplied to normal operating theatres. The patients were nursed directly beneath the canopies in the unidirectional airflow. The air supply pressurized the rooms and prevented the ingress of air from the surrounding side wards and corridors. The temperature and humidity of the supply air in these rooms could be controlled within wide limits.

Either of two sophisticated airbeds was used; large peripheral burns were initially treated on a hover-bed, similar to an inverted hovercraft, where the patient was nursed on a cushion of air[18]. A flexible fabric acted as a seal between the bed and the skin to control the amount of air escaping past the patient. This bed is shown in Figure 11.3(a) positioned beneath one of the linear airflow canopies. The patient was virtually levitated and no part of the body was in firm contact with any other surface; the moving air dried the injured area and produced a stable eschar, or scab, much quicker than if the wound was left to dry naturally (Figure 11.3(b)). The speedy formation of this eschar reduced the loss of body fluid through the burned area and prevented the ingress of potentially pathogenic micro-organisms. There was an even air pressure over the whole body so the patient was far less uncomfortable than if he were on a conventional bed.

The volumes of air required for this bed were considerable and up to 25 m³/min with the temperature and humidity accurately controlled.

The second stage was to nurse the patient on a 'low air loss bed'[19] which consisted of 21 air sacks arranged in several groups; each group of sacks could be set at a given pressure so the patient was evenly supported. The air supply to the bed was temperature controlled and the fabric of the air sacks was water vapour permeable, so that any burn exudate and water lost from the skin would not collect as a film but would be diffused in the air circulating within the sacks. This air bed is shown in Figure 11.4.

Nursing patients on these airbeds beneath the linear downflowing ultra-clean airstreams, involved some unusual comfort and thermal balance problems, particularly when a patient was nursed on a hover-bed. The rapid air movement evaporated a large volume of exudate from the burn to give a high rate of cooling. The air temperature had to be raised and the humidity increased to keep the patient comfortable and it was not uncommon for a patient to require an air temperature of 45°C at 60 per cent relative humidity. Temperatures lower than 40°C with humidities below 50 per cent produced shivering and an intolerable feeling of cold. On occasions when dry bulb temperatures higher than 40°C were not available it was necessary to increase the relative humidity of the air to 65–70 per cent to make the conditions tolerable. It is interesting to compare these 'comfort' temperatures with the temperature of about 45°C of still air which will feel uncomfortably hot.

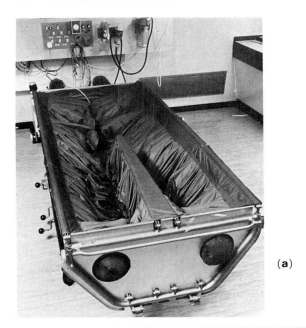

(a)

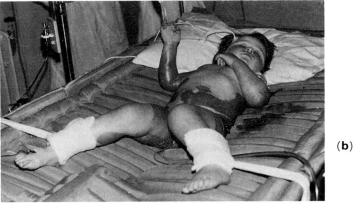

(b)

Fig. 11.3 (a) The 'high air loss' or 'hover-bed' used in the treatment of major burn injury. The patient was supported on a cushion of air with the flexible fabric acting as an air seal at the body surface (b).

As well as nursing badly burned patients on the hover bed, trials were conducted to determine the responses of 'normal' subjects who did not, of course, have the enormous water loss and consequent evaporation of the burned patients. The comfort temperatures were between 39 and 41°C at relative humidities of 40–50 per cent for a series of eight subjects.

In all of the burned patients and the normal subjects nursed on the hover bed, measurements were taken of skin temperature at three sites, together

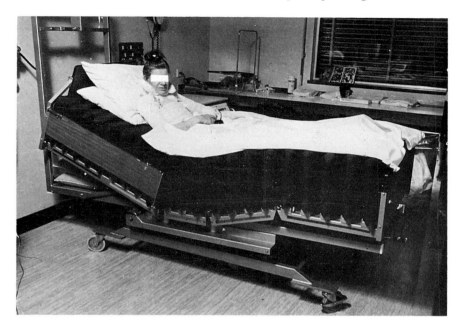

Fig. 11.4 A 'low air loss bed' with the patient supported on air-filled sacks in which the air pressure and temperature can be adjusted to the requirements of the individual. Manufactured by Mediscus Products Limited, Wareham, England.

with deep body temperatures using rectal probes. In a burned patient these temperatures and measurements of environmental air temperature and humidity, were recorded continuously by means of chart recorders during the first two to three days of treatment, so a continuous assessment of rectal temperatures could be made in relation to the air temperature and humidity supplied to the hover-beds. It was these two environmental variables that were used to 'drive' the bed in an attempt to maintain patient 'comfort' and rectal temperatures within desired limits. The 'comfort' conditions of around 45°C also maintained constant rectal temperatures.

There were a number of practical nursing management difficulties concerned with the hover-beds which, because of the large amounts of air passing through them, were extremely noisy and had a strange appearance which caused anxiety and fear in some patients. The considerable expertise needed to run them was not always available and, eventually, treatment of major burns by exposure was superseded by other methods. The hover-bed undoubtedly had some novel features and was effective; there may well be a revival of this technique.

The low air loss bed (LALB) proved to be a great success both in preventing pressure sores and providing a comfortable temperature controlled environment for the burned patient. It had characteristics which helped in the difficult task of handling these patients particularly for the lengthy process of changing extensive dressings. This bed has since become popular for a wider range of treatments including long term use by immobile patients in their own homes.

The air ventilation systems and their sophisticated controls enabled a number of studies to be carried out on thermal stress and bacteriological cleanliness and these are summarized in the following sections.

Air movement and thermal stress

In Chapter 9 the relationship between the sensation of comfort and environmental temperature is examined in detail. The study of the relationship between the actual heat transfer from the human body and the sensation of comfort is complementary to the subjective assessment of the environment by various questionnaire and survey techniques.

At the Mount Vernon Burns Unit, using a model to represent the human body, the convective cooling produced by the vertical downflowing airstreams in the intensive care rooms was measured. Cooling coefficients were assessed at two different downflow velocities and were compared to results of similar tests in rooms without ventilation. The model used was an elliptical cylinder having a total surface area of $1.8\,\text{m}^2$. It was fitted with electrical heaters deriving power from a control console, the power being measured in terms of the current and voltage supplied. Thermocouple thermometers were used to measure the temperatures of the model surface and of the surrounding air.

The model could be positioned either vertically or horizontally and was used in both these configurations beneath the linear airflow canopies. Experiments were carried out to determine the convective cooling coefficients with the model freely suspended horizontally and also with it resting on an ordinary hospital bed and on the low air loss bed (LALB), either with the cylinder completely uncovered or with it covered by a sheet to represent the degree of bedding generally used within the burns unit.

The shape of the convective boundary layer flow (Chapter 4) over a standing subject is quite different from that found in the prone or supine posture. These flow patterns were considerably modified by the downflowing airstreams of the ultra-clean air ventilation systems in the burns unit and this is

Table 11.2 Convective cooling coefficients for a horizontal and vertical cylindrical model in 'still' air and when exposed to linear downflowing airstreams at two different velocities.

	Convective cooling coefficient h_c ($\text{W/m}^2.°\text{C}$)		
	'Still' air	Downflow 0.30 m/s	Downflow 0.65 m/s
Vertical cylinder	6.7	5.9	4.5
Horizontal cylinder	6.6	5.7	10.8

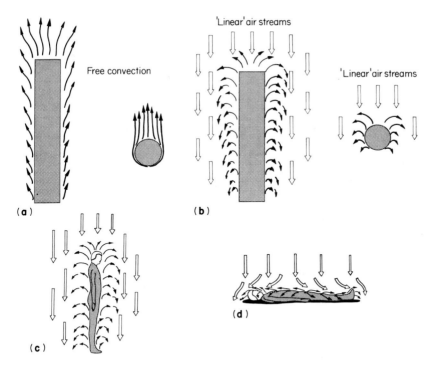

Fig. 11.5 Diagrams showing the modifications to the convective upflows by downflowing airstreams for a heated cylindrical model used to measure heat transfer coefficients ((a) and (b)) and for a human subject ((c) and (d)) in different postures.

illustrated diagrammatically in Figure 11.5 and the relationships between cooling coefficient, posture and speed of the linear downflowing airstreams are summarized in Table 11.2. When the model was vertical in the downflowing airstreams the convective cooling coefficient was lower than in the 'still' air situation. The downflow was effectively 'in competition' with vigorous convective upflow generated over the whole height of the cylinder. The interaction of the two airstreams slowed down and thickened the boundary layer, so the temperature gradients were less steep with a consequent reduction in the convective heat loss. The downflowing airstreams were not fast enough to blow away the natural convective flow altogether and produce conditions of true forced convection. However, with the cylinder horizontal, the situation was different; when the downflow was 0.3 m/s the cooling coefficient was lower than in 'still' air but when the velocity of the downflow was 0.65 m/s the up-flowing convective currents were displaced to produce conditions of forced convection which resulted in a large increase in the convective cooling coefficient.

The results showed that if the linear downflowing air speed was too high, the patient, lying on the bed, was at a thermal disadvantage compared with

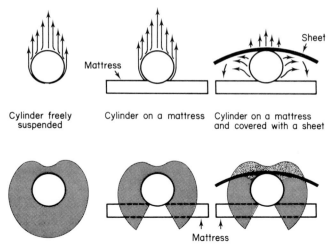

Cylinder freely
suspended

Cylinder on a mattress

Cylinder on a mattress
and covered with a sheet

$h_0 = 12.8$ W m^{-2} K^{-1} $h_0 = 9.4$ W m^{-2} K^{-1} $h_0 = 6.1$ W m^{-2} K^{-1}

Fig. 11.6 Diagrams of air flows and convective cooling coefficient 'envelopes' for a horizontal cylinder in three configurations, h_0 denotes the overall heat loss coefficient. The arrows in the top part of the diagram indicate the direction and thickness of the convective airstreams generated around the cylinder and the lower part shows the associated convective heat loss 'envelope' where the radial distance through the dotted area to the cylinder circumference is proportional to the local convective heat loss at each point on the cylinder surface. The effect of the mattress is to remove part of the convective envelope and the sheet further modifies the area of envelope indicated by the larger dots.[33]

attendant nursing staff who were standing. This turned out to be an important factor in determining the correct air temperature for the treatment of patients since with high airspeeds the increased cooling necessitated higher dry bulb air temperature for thermal equilibrium to be maintained.

Figure 11.6 attempts to represent the complex relationship between the free convective airstreams and heat loss coefficients for the heated cylinder, either freely suspended in air, or placed on a mattress and covered or uncovered by a sheet. The effect of the linear downflowing airstreams on the cylinder in these situations is too complicated to illustrate adequately but Table 11.3 summarizes the coefficients measured in these experiments.

The studies at Mount Vernon showed that comfort for patients and staff could be prejudiced by air ventilation rates set too high even though such air movements may have been theoretically desirable from a microbiological point of view. However, when the airflows from the overhead canopies in the burns unit were reduced to 0.3 m/s it was found, over a period of several months, that bacterial cleanliness was substantially the same as at the higher airflow rate of 0.65 m/s. A downflowing airspeed of about 0.3 m/s would seem to be the maximum that should be used in such air-conditioned areas in order to avoid the risk of exposing a patient to excessive heat loss. The airflows in the intensive care rooms were set at this lower level, with a great saving of

Table 11.3 Convective cooling coefficients for a horizontal cylindrical model freely suspended and when placed on a bed in 'still' air and when exposed to linear downflows at two different velocities. H_c represents total heat loss by convection, radiation and conduction because of the difficulty in separating the various modes with the cylinder on a mattress.

	Overall cooling coefficient H_c (W/m².°C)		
	'Still' air	Downflow 0.3 m/s	Downflow 0.65 m/s
Horizontal cylinder freely suspended in air	12.8	11.0	17.4
Horizontal cylinder on a bed uncovered	9.4	13.3	12.5
Horizontal cylinder on a bed, covered with sheet	6.1	9.1	5.8

fuel costs to heat the air and of maintenance costs for the high efficiency filters.

Members of the burns research team slept on ordinary hospital beds beneath the linear downflowing airstreams with air velocity set at 0.65 m/s. It was an uncomfortable experience as the high airflows cooled exposed parts, such as the hands and face, producing large temperature differences between uncovered and covered areas of the body. To warm the extremities, dry bulb temperatures were raised, but the covered parts of the body were then too warm. Respiratory discomfort was noted by several people who slept in these rooms, and there were complaints of excessive drying of the nose and mouth by the following morning.

The noise of the air systems and the general awareness of being in a moving airstream also added to the unpleasantness of the environment. None of these criticisms were echoed by patients who were nursed beneath these canopies; they were probably fully occupied with the consequences of their serious injuries and would not expect the same degree of comfort that they might have enjoyed in their own homes.

The air-conditioning systems in the intensive care rooms at Mount Vernon were similar to those installed in a number of operating theatres. However, the fact that burned patients were often exposed to this environment for many weeks, or even months, presented problems not generally encountered in the operating room where the patient and surgical team remain for only a short time, some hours at the most.

Within the intensive care rooms at the burns unit, there were two hand operated controls, one to change dry bulb air temperature and the other to alter the relative humidity of the air supply. The provision of two controls led to some rather unusual temperature management problems because the dry bulb (DB) temperature and the relative humidity (RH) of the downflowing air could be controlled independently. The problems that arose can be illustrated as follows.

If a patient was cold, then it was possible either

1. to increase DB temperature and leave RH or
2. leave DB and increase RH or
3. increase both DB and RH.
   It was also possible to achieve thermal comfort by:
4. reducing DB and increasing RH or
5. increasing DB and reducing RH.

Such a large number of choices was found to be just acceptable while the unit was staffed by a research team. However, on the departure of this group and even with the unit staffed by highly-trained nurses, such a large choice of action was found to be unworkable. It was eventually found necessary to limit the environmental control to dry bulb (DB) temperature alone. A possible complicating factor may have been that one variable was the relative humidity of the air which was difficult to relate to the requirements of the patient.

Air systems in laboratories

In the last few years there has been an increasing awareness of the potential dangers associated with material handled in chemical and microbiological laboratories. Much of this concern has stemmed from outbreaks of smallpox which have been traced to research laboratories which had been considered secure[20, 21]. There is now a considerable industry in the supply and installation of laboratory equipment designed to contain infectious material. Codes of practice for laboratory procedures together with various standards for containment facilities[22, 23] have been published which seek to improve conditions in laboratories generally.

The design and testing of chemical fume cupboards and microbiological safety cabinets play a significant part in the provision of safe laboratory facilities. A number of studies have shown that much existing apparatus in laboratories is sub-standard[24, 25, 26, 27] and requires upgrading. It is now necessary to introduce designs for containment facilities having proven performance, but it is equally important that such equipment is compatible with other laboratory facilities, in particular with heating and ventilating systems. The principles of containment and the thermal requirements of laboratory workers are sometimes difficult to reconcile. Careful design is particularly necessary to achieve thermal comfort in the presence of large volumes of moving air. In systems where the air is eventually discarded, heating costs can be prohibitively high unless energy conservation methods are employed.

There are laboratory areas where specific problems exist, for example, laboratory suites for handling viruses such as those that cause Lassa fever and smallpox and rooms suitable for genetic manipulation, radiopharmaceutical and carcinogen work. The combined skills of people such as micro-biologists, chemists, architects, air-conditioning engineers, physiologists, physicists and administrators are needed to rationalize future designs of such facilities. At present there are no generally accepted design performance criteria for complete containment laboratories and this is an area where considerable research effort is needed if disastrous laboratory accidents are to be avoided.

The neonatal microenvironment

The microclimate of a newborn infant nursed naked within a baby incubator is exceptional in several respects, and deserves special consideration. Premature infants may have deficiencies in temperature regulation by having limited sweating responses. Such babies require the warm conditions of an incubator to minimize their energy loss and avoid problems of thermal regulation. For whatever reasons infants are nursed in incubators, it is the only place where they are nursed naked for long periods of time. Because of this, a study of the neonatal microclimate is important, both from the point of view of basic thermal physiology and also because it is essential that the environment within an incubator is well understood and provides acceptable thermal conditions that do not stress the infant[28].

Studies have been made to define the neonatal microclimate. Colour infra-red thermography has been used to visualize the skin surface temperatures over infants nursed in a thermoneutral environment[29] so a mean value for the whole body surface temperature could be obtained which, with the knowledge of environmental air and surface temperatures, and air movement, allowed the heat transfer with the surroundings to be determined. The environmental conditions within an incubator could then be chosen so that the infant experienced conditions where energy loss was minimal and where the body would be in thermal equilibrium with the surroundings. In these studies a number of infants were visualized and the temperature distributions were substantially similar. The pattern was much less structured than in the child or adult with relatively few features due to the direct influence of subcutaneous thermogenic or insulating structures. The thermoneutral temperature for the infants was around 32°C and when the incubator air temperature was cooled to around 27°C subtle skin temperature changes occurred with structures underlying the skin surface becoming visible as seen in Plate 15(a).

In this plate the anterior view shows two 'hot spots' overlying the jugular vein and the carotid artery region, and in the posterior view a 'Y' shaped cooler area is just discernible in the intrascapular region extending downwards. It was considered unethical to expose the infant to temperatures below 27°C but it may be expected that in cooler conditions the effect of the structures underlying the skin would become more dominant in determining the temperature distribution pattern, in much the same way as in the adult.

In a further study, skin temperature changes (similar to those observed in adults and described in Chapter 2) were found over neonates. An example is shown in Plate 15(b) where thermograms were recorded at regular intervals during an 80-minute period. This plate also shows a computerized version of

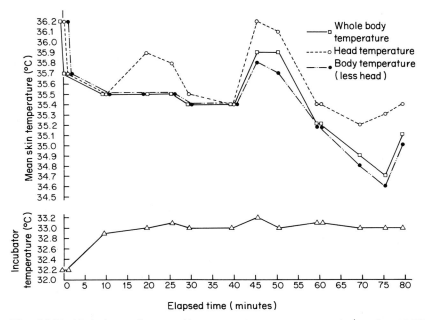

Fig. 11.7 Variations of mean skin temperature over a neonate in a temperature controlled incubator (lower curve shows incubator air temperature).

one of the images to illustrate the evaluation of mean temperatures for the whole body, head and trunk. The results of this analysis are shown in Figure 11.7 where these temperatures are plotted separately; also shown is the incubator environmental temperature.

A rhythm in skin temperatures is suggested with the head perhaps slightly dominant. In all of these studies the head was not found to be significantly or consistently at a higher temperature than the trunk despite the very high metabolism in the brain (relative to the rest of the body). This is probably because there is very efficient heat exchange between the brain and the other body organs and that, relatively, the head has a large surface area (some 20 per cent of total skin area) which enables a high proportion of total heat loss to the environment to occur from the head.

The infant temperature distributions were used to determine areas of skin which would characterize the overall mean surface temperature. The identification of such a site is important when controlling incubators from a reference area on the skin surface. The site must be responsive to changes in environmental conditions. Overall mean skin temperatures obtained from analysis of infra-red thermograms were compared with those obtained from a number of spot measurements with thermocouple probes; the results suggested that the upper arm, the thigh and the buttocks were all reasonably near to the mean temperature and that they were regions which were also responsive to environmental change. Any of these areas would therefore be suitable for attaching a sensor which could control the air temperature within the incubator canopy.

The calculation of overall skin temperature depends upon the weighting given to each spot measurement. The weightings for adults and infants are different; in the infant the arms and legs constitute some 45 per cent of the total area and the head accounts for over 20 per cent. In the adults, the proportions are different with the head being only some 7 per cent of the total body surface area. This is illustrated in Figure 11.8 which shows the difference in relative area, as a percentage of the total area, for the infant and the adult.

The neonatal convective boundary layer

The temperature gradient between the skin surface and the infant's surroundings enables heat transfer by conduction, convection, radiation and evaporation to occur. The convective microclimate around the infant is generated in exactly the same way as for the adult; the air in immediate contact with the skin surface becomes more buoyant than the general surrounding air and produces a convective boundary layer flow over the baby. This flow has been visualized using the Schlieren method[30] and for this investigation an infant incubator was modified by having special pieces of optical glass inserted into apertures in two sides of the incubator canopy; this enabled the Schlieren beam to pass through the canopy and reveal the convective microclimate generated by the infant.

For many years the most important criteria for environmental control within the incubator canopy were to maintain the air and wall temperatures

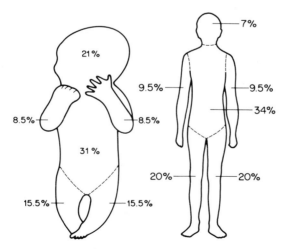

Fig. 11.8 The regional surface areas of the neonate (Klein and Scammon, (1960) *Proceedings of the Society of Experimental Biology* **27**, 463–6) compared with those of the adult (DuBois (1916) *Archives of Internal Medicine* **863**, 101–3). A major proportion of changes which occur with growth are those of the head, becoming smaller with age and the legs where the relative area increases.

as close as possible to a pre-set level. It was also part of the design that there should be the smallest air temperature variation possible throughout the canopy. Control systems have progressively improved and it is now possible to have such temperature variation as low as 0.1°C. This precise control has been achieved mainly by moving the air rapidly to ensure a high degree of turbulent mixing of the air. In general, the greater the degree of mixing within the canopy, the less spatial variation there is in air temperature. However, it is not generally appreciated that by increasing air movement the infant is subjected to greater convective cooling. Thus the air temperature must be raised if comfort is to be preserved and thermal stress avoided.

The Schlieren investigation revealed the 'natural' thermal currents generated by the infant itself and showed how these airstreams were modified by the ventilating airflows within the incubator. At an air temperature of 32°C with a mean skin temperature of 36°C, in the incubator canopy without any ventilating air movement the convective air currents stream upwards away from the whole skin surface with a maximum speed of about 0.05–0.08 m/s; the total volume of air moving over the top of the infant is about 1 litre/sec. The Schlieren method clearly demonstrates the way in which expired air is carried upwards as it leaves the infant's nose and mouth. In 'still' air these natural convective airstreams carry away about 20 per cent of the total heat production from the infant; the rest of the heat is lost in the form of radiation (60 per cent), conduction and evaporation. There is a greater conductive heat loss from the baby than the adult because a large proportion of the body surface is in contact with the mattress.

The situation where there is no airflow within the incubator canopy is highly artificial. Many incubators in use have an airflow with a maximum velocity of 0.2–0.25 m/s. This level of environmental air movement is regarded as being, for the adult, 'still' air conditions. To the infant, however, it is far from 'still', and when the incubator is turned on the Schlieren system shows that the baby may be in conditions that can fairly be described as a 'howling' gale. The recirculation and turbulent mixing within the canopy completely negates the natural convective upflow and the infant loses heat by forced convection. Incubator ventilation systems have failed to account for this effect on the baby's natural microclimate and designs have almost entirely been concerned with maintaining the homogeneity of the temperature throughout the incubator canopy.

Assessments were made on the cooling that a baby would experience within the incubator, using the Kata thermometer (see Chapter 4) to determine the cooling power of the environment both with the incubator switched on and also with no airflow within the canopy. The results showed that the baby would lose 25 per cent more heat when the ventilating airflows were on than in 'still' air at the same temperature. This means that the convective part of the heat loss was increased by some 60 per cent by the ventilation airflows.

In the clinical management of infants nursed in incubators it has been customary to determine the thermo-neutral temperature, that is the temperature at which the incubator air should be set, from considerations of the infant's size, gestational age and oxygen consumption. These calculations were based on work by Hey[31] who studied the thermal balance of infants within a carefully controlled calorimeter. The air movements within this calorimeter were much slower than found in infant incubators used in hospital wards, however the results of these studies were applied directly to the ward situation without taking account of increased air movement in particular designs of incubator, but only the air temperature.

In the light of the Schlieren observations, further investigations were made with incubators having slower airflows which did not negate the natural convective airstreams but mixed with them, so that neither airflow dominated. It was found that homogeneity of temperature throughout the canopy could be satisfactorily maintained at these low airspeeds and the results of this work have now been applied to the design of new infant incubators that are beginning to appear in hospital wards.

References

1.   Bernal, J. D., (1966). Quoted by Goldsmith, M. and MacKay, A. In: the Introduction to *The Science of Science*. Penguin, London.
2.   McIntyre, D. A. (1981). Design requirements for a comfortable environment. In: *Bioengineering, Thermal Physiology and Comfort* pp. 195–220. Ed. by K. Cena and J. A. Clark. Elsevier Scientific, Amsterdam, Oxford, New York.
3.   Auliciems, A. (1969). Thermal requirements of secondary schoolchildren in winter. *Journal of Hygiene, Cambridge* **67,** 59.
4.   O'Sullivan, P. E. and Cole, R. J. (1973). The thermal performance of school buildings. *Symposium 'Environmental Research in Real Buildings'*. Personal communication.
5.   Longmore, J. and Ne'eman, E. (1974). The availability of sunshine and human requirements for sunlight in buildings. Symposium on Environmental Research in Real Buildings. *Journal of Architectural Research* **3,** 24–9.
6.   Ne'eman, E. (1974). Visual aspects of sunlight in buildings. *Lighting Research and Technology* **6,** 159–64.
7.   Chandler, E. J. (1978). The man-modified climate of towns. In : *The Built Environment*. Ed. by J. Wernham and W. W. Fletcher. Blackie, Glasgow and London.
8.   Marcos-Assaad, F. N. (1978). Design and building for a tropical environment. In: *The Built Environment* pp. 26–81. Ed. by J. Wernham and W. W. Fletcher. Blackie, Glasgow and London.
9.   *Proceedings of the Fourth Biennial Infra-red Information Exchange* (August 1978). St. Louis, Missouri. Published by AGA Corporation 550 County Av. Secaucus, N.J. 07094.
10.   Broadbent, G. (1973). *Design in Architecture. Architecture and the Human Sciences*. J. Wiley, London and New York.
11.   Bissell, M., Diamond, M. C., Ellman, G. L., Krueger, A. P., Orenberg, E. K. and Sigel, S. S. (1981). Air Ion research. *Science* **211,** 1114.
12.   The General Board of Health. (1857). *Report of the commission appointed to enquire into the warming and ventilation of dwellings.*
13.   Bedford, T. (1936). *The warmth factor in comfort at work*. Industrial Health Research Board, Report No. 76.
14.   Report of a joint working party of the Department of Health and Social Security, the Medical Research Council and the Regional Engineers Association. (1972). *Ventilation in operating suites*. Available from the MRC, 20 Park Crescent, London W1N 4AL.
15.   Lidwell, O. M., Lowbury, E. J. L., Whyte, W., Blowers, R., Stanley, S. J. and

Lowe, D. (1982). Effect of ultraclean air in operating rooms on deep sepsis in the joint after total hip or knee replacement: a randomised study. *British Medical Journal* **285,** 10.

16.   Morris, R. H. and Wilkey, B. R. (1970). The effects of ambient temperature on patient temperature during surgery not involving body cavities. *Anaesthesiology* **32,** part 2, p. 102.

17.   Morris, R. H. (1971). Influence of ambient temperature on patient temperature during intra-abdominal surgery. *Annals of Surgery* **173,** 230.

18.   Scales, J. T., Hopkins, L. A., Bloch, M., Towers, A. G. and Muir, I. F. K. (1967). Levitation in the treatment of large-area burns. *Lancet* **i,** 1235–40.

19.   Scales, J. T. and Hopkins, L. A. (1971). Patient-support system using low-pressure air. *Lancet* **ii,** 885–8.

20.   Department of Health and Social Security (1975). *Memorandum on the control of outbreaks of smallpox.* HMSO, London.

21.   Department of Health and Social Security (1978). *Investigation into the 1978 Birmingham smallpox occurrence.* HMSO, London.

22.   Department of Health and Social Security (1976). *Control of laboratory pathogens very dangerous to humans.* HMSO, London.

23.   Department of Health and Social Security (1978). *Code of practice for the prevention of infection in clinical laboratories and post-mortem rooms.* HMSO, London.

24.   Clark, R. P. and Mullan, B. J. (1978). Airflows in and around linear downflow safety cabinets. *Journal of Applied Bacteriology* **45,** 131–6.

25.   Hughes, D. (1980). *A literature survey and design study of fume cupboards and fume-dispersal systems.* Occupational Hygiene Monograph No. 4. Science Reviews Ltd., Northwood, Middlesex.

26.   Clark, R. P. (1983). *The performance, installation, testing and limitations of microbiological safety cabinets.* Occupational Hygiene Monograph No. 9. Science Reviews Ltd., Northwood, Middlesex.

27.   Clark, R. P. (1982). The performance of containment facilities. *Journal of the Society of Environmental Engineers* **24,** 31–5.

28.   Stothers, J. K. (1984). Special thermal problems of the neonate. In: *Recent Advances in Medical Thermology.* Ed E.F.J. Ring and B. Phillips. Plenum, New York.

29.   Clark, R. P. and Stothers, J. K. (1980). Neonatal Skin Temperature distribution using infra-red colour thermography. *Journal of Physiology* **302,** 323–33.

30.   Clark, R. P., Cross, K. W., Goff, M. R., Mullan, B. J., Stothers, J. K. and Warner, Ruth M. (1978). Neonatal natural and forced convection. *Journal of Physiology* **284,** 22–3P.

31.   Hey, E. N. (1969). The relationship between environmental temperature and oxygen consumption in the newborn baby. *Journal of Physiology* **200,** 589–603.

32.   Johnston, Ivan, D. A. and Hunter, Andrew, R. (1984). *The Design and Utilization of Operating Theatres.* Edward Arnold (Publishers) Ltd, London.

33.   *Thermal Physiology and Comfort* (1981). Ed K. Cena and J. A. Clark. Elsevier, Amsterdam and New York.

Index